Boel Berner
Strange Blood

Medical Humanities | Volume 5

Boel Berner is a sociologist, historian, and professor emerita at Linköping University in Sweden. In her research she investigates the character and power of expertise, historically and today. She has studied education and work, the gendered nature of technical knowledge, household modernization, and issues of risk. Her current work is oriented towards the history of medicine. It focuses, besides questions of blood donation and transfusion, on the politics of blood group analysis in the interwar years.

Boel Berner
Strange Blood
The Rise and Fall of Lamb Blood Transfusion
in 19th Century Medicine and Beyond

[transcript]

Bibliographic information published by the Deutsche Nationalbibliothek
The Deutsche Nationalbibliothek lists this publication in the Deutsche Nationalbibliografie; detailed bibliographic data are available in the Internet at http://dnb.d-nb.de

This work is licensed under the Creative Commons Attribution-NoDerivatives 4.0 (BY-ND) license, which means that the text may be shared and redistributed, provided credit is given to the author, but may not be remixed, transformed or build upon. For details go to
http://creativecommons.org/licenses/by-nd/4.0/
To create an adaptation, translation, or derivative of the original work, further permission is required and can be obtained by contacting rights@transcript-verlag.de

Creative Commons license terms for re-use do not apply to any content (such as graphs, figures, photos, excerpts, etc.) not original to the Open Access publication and further permission may be required from the rights holder. The obligation to research and clear permission lies solely with the party re-using the material.

© 2020 transcript Verlag, Bielefeld

All rights reserved. No part of this book may be reprinted or reproduced or utilized in any form or by any electronic, mechanical, or other means, now known or hereafter invented, including photocopying and recording, or in any information storage or retrieval system, without permission in writing from the publisher.

Cover layout: Maria Arndt, Bielefeld
Cover illustration: Francisco de Zurbarán, Agnus Dei (1635-1640), © Museo Nacional del Prado
Printed by Majuskel Medienproduktion GmbH, Wetzlar
Print-ISBN 978-3-8376-5163-8
PDF-ISBN 978-3-8394-5163-2
https://doi.org/10.14361/9783839451632

Printed on permanent acid-free text paper.

Contents

Prologue 9

Introduction: 'The mighty influence of strange blood' 11

PART I: SETTING THE SCENE

1. Using the blood of others 17
   The beginnings 18
   The return... 21
   Direct or indirect transfusion? 23
   Gaining acceptance 25
   The revival of lamb blood transfusion 28

2. Ambitions and connections 31
   The sanguine local doctor 31
   The polemicist.. 34
   The entrepreneur 36
   The context.. 40

PART II: PRACTICES

3. Blood on the battlefield 45
   Wars, wars, wars 46
   War-time modernization 49
   Using Roussel's apparatus – or not? 51
   The animal blood alternative... 53

Saving the apparently dead 55
War and medical innovation... 58

4. Blood for the lungs 61
Consumption challenges 62
The benefits of lamb blood 66
Performing transfusions 67
Experiencing lamb blood transfusion.. 72
Getting better?... 74
Still worth trying? 77

5. Asylum experiments 81
Pellagrous conditions 82
Testing transfusion.. 86
Transfusion and the Risorgimento of Italian science............................... 88
First experiences 90
A transfusion competition 95
The Brescia experiment 96
Understanding improvement..99
Assessing experiments 101

PART III: CONTROVERSY

6. Proofs and refutations... 105
Bedside medicine 108
Hospital medicine.. 111
Laboratory medicine 114
Laboratory experiments contested 117
Clinical experience contested 120
The problem with statistics 122

7. Transgressions 127
Using animals 128
Crossing boundaries 133
Accepting transgression 134
Was it worth it?..137
Overstepping boundaries 139

PART IV: THE FALL

8. Winding up 145
The condemnation 147
Understanding blood 148
Lessons learnt 151
Human trajectories 153
No more blood on the battlefield? 156

Epilogue: The return 159
The advent of serotherapy 160
Hasse vindicated? 162
French interventions 164

Notes 167

Sources and Literature 183
Archives 183
Websites 183
Literature 183

Acknowledgements 203

Index of Places 205

Index of Names 209

Prologue

It is late May 1873, springtime in the Harz Mountains in northern Germany. Flowers are in bloom, birds are singing, but in the home of Heinrich Krüger, a cattle dealer in the village of Schwenda, there is sorrow and despair. Thirteen-year old Hermine Krüger has suffered from diphtheria since the autumn of 1872. Now her condition is deteriorating. She has severe abdominal pain, a forced and wheezing respiration, strong sweating, no appetite.

Oscar Hasse, a well-known doctor from the town of Nordhausen some 40 km away, is called for. Upon arrival, he finds Hermine in bed, with a pale, bloated appearance, short, groaning breath, very weak and rapid pulse. It turns out that she also suffers from a severe bronchial catarrh. She is so weak that, to sit, she has to be supported by several persons – she cannot keep her head up. The various tonics prescribed have been of no use. She seems beyond salvation.

Upon the urgent demand of the family, Hasse decides to perform a blood transfusion but not, as he has previously done, with human blood. The girl's relatives are just too weak themselves. Instead, a strong six-month-old lamb is procured. Hasse ties it to a board, exposes its carotid artery, and closes it temporarily with a clamp. He then inserts a glass tube into the vessel and connects it to a rubber tube containing a carbonate of soda solution. A second glass tube is inserted into the *vena mediana* of the sick child. The lamb is brought into the sick room and laid next to the patient with its neck close to the girl's arm. Hasse unfastens the clamp and lets the lamb's blood push out the soda solution. He then swiftly connects the two cannulas with the rubber tube. Propelled by the heartbeat of the lamb, its blood now flows into the child.

And behold! The transfusion is an ordeal, but the girl's forces return. She sleeps well, eats with an appetite, and has no more stomach pains. Hasse brings her to Nordhausen for treatment with gymnastics and galvanic currents to strengthen her muscles after almost a year in bed. By the end of the summer, Hermine Krüger seems fully restored.[1]

Introduction: 'The mighty influence of strange blood'

The experience of a small-town German doctor would, in the mid-1870s, start a formidable transfusion craze. Oscar Hasse of Nordhausen am Harz tried transfusion with lamb blood on fifteen patients, reported positive results in meetings and publications, and suddenly hundreds of lamb blood transfusions were made in clinics, hospitals and lunatic asylums across Europe and the USA. 'The blood of lamb and sheep was flowing in streams, the literature on transfusion was growing like an avalanche from day to day', a contemporary observer noted.[1] Doctors used it as a cure for phthisis, pellagra, cancer and epilepsy, suggested it as a means to reawaken seemingly dead soldiers on the battlefield. It was seen as 'life-giving, despite its repulsive animality'.[2]

I first encountered this phenomenon when researching for a book on the history of Swedish blood transfusions. I found that several lamb blood transfusions had been made in Sweden in the 1870s.[3] It turned out that they were part of a wider international story. Lamb blood transfusion appeared in the early 1870s, caught on and multiplied, then disappeared. This piqued my curiosity. Why this sudden fervour for transfusing strange blood? How was it undertaken, by whom, and how did the patients feel? And, most importantly: Did it work?

This book will give some answers. It is the result of digging into archives, consulting esoteric documents, and visiting hospitals and universities across Europe. It will be rich in details about mid-19th century sick bed encounters and laboratory experiments; it will show hope and disappointment, human and animal suffering alike. Geographically, we move from North Carolina to St. Petersburg, from central London to the German countryside, from southern Sweden to northern Italy. We get to know the ambitions of the main actors, the experimental nature of the intervention, and its international ramifications.

We also follow the conflicts between proponents of clinical experience and scientific proof. This was a controversial therapy, hotly debated at the time. Strangely, it has been almost totally neglected by historians of medicine and science. If taken up at all, lamb blood transfusion has been dismissed as a roadblock to medical progress, a thoughtless experiment with patients as unsuspecting guinea pigs. My perspective is different. I think that investigating a 'losing' practice like lamb blood transfusion reveals, as medical historian Anita Guerrini suggests, that '[f]ew things are simply right or wrong, either ethically or scientifically. More often they are a muddle of mixed motives and half-clear ideas.'[4] I therefore take seriously the arguments of 19th century doctors, patients and scientists, try to understand their mixed motives and muddled ideas, and situate them in a larger context of professional ambitions and uncertainties. My story stays close to what happened across Europe and the USA when physicians tried to save their patients' lives or sanity with the pulsating blood of a lamb. Their accounts take us into a world of pain where the patient sweats and coughs, the bleeding does not cease, and the breath becomes weaker with every second. What to do? Where to turn? Why not try the remedy that the medical press says works wonders with the severely ill and test, as one American doctor phrased it, 'the mighty influence of strange blood'?[5]

Lamb blood transfusion was, in many ways, a transgression. Its use broke rules and exceeded boundaries. To some physicians in the 1870s, it was a daring, but not irrational, choice when previous treatments had failed. To others, it was a dangerous experiment on helpless patients, and a return to pre-modern ideas about the occult qualities of blood. Thus, the experiment with lamb blood transfusion was a political phenomenon. It upset medical hierarchies and truths. It challenged medical knowledge, ethics and expertise, gave rise to controversy and debate. It had ramifications also outside the medical world. The rise – and fall – of lamb blood transfusion was, I will show, linked to mid- and late-19th century struggles for national revival, social justice and military advance.

***

The book proceeds as follows.[6] Part I sets the scene. Chapter 1 traces the often-interrupted history of blood transfusion, from its beginnings with animal blood in the 1660s, through the abandonment of the therapy in the centuries thereafter, and all the way to its revival with human blood in the early 19th century. Then, in the 1870s, came the unexpected return of lamb blood transfusion. I will give a background to the excitement and confusion that would follow.

Chapter 2 presents three ambitious men – Oscar Hasse, Franz Gesellius and Joseph-Antoine Roussel – who more than others were implicated in what was seen as a transfusion 'epidemic' in the 1870s. Very different in personality, they shared a certain outsider position vis-à-vis established medical hierarchies. Still, they managed to put a mark on the history of transfusion. They met with acclaim and disdain, success and failure. We follow their respective trajectories up to and including the crucial year of 1874 when the enthusiasm for transfusion (with both human and animal blood) was at its peak. They will then reappear as central actors in several other chapters. I will eventually reveal what they did when blood transfusion was no longer in vogue.

Part II of the book – Practices – takes us into three, quite dissimilar medical worlds where lamb blood transfusion was advocated as a panacea and/or was practiced (with varying results). In chapter 3 we follow doctors onto the bloody battlefields of mid-19th century wars. In chapter 4, we move to the more serene settings of homes and hospitals where tuberculosis patients coughed their lungs out, and in chapter 5, we share the chaos and despair of mid-19th century Italian asylums where pellagra sufferers awaited a certain death. In many instances, lamb blood transfusion was argued for and sometimes it was tried out. Why was it done? How did doctors go about finding a suitable lamb, preparing it and connecting it to a patient, and how did their patients react to the sudden influx of strange blood into their veins? These situations I will depict in careful and sometimes gory detail.

But did it work? Part III is about controversy. There seemed to be no definite proof for or against the healing powers of lamb blood transfusion, neither from clinical experience nor from animal experiments. In chapter 6, I disentangle the somewhat confusing debate. I link the arguments and counterarguments to relations of power between doctors in clinical settings and physiologists in their laboratories. They had different ways of approaching diagnoses and cures, assessing evidence and results. Still, and as we shall see in chapter 7, both clinical and laboratory practices involved transgressions. Humans and animals were experimented on in often painful and perhaps futile ways. By what right did doctors and scientists tamper with the lives of others? How did they negotiate the boundaries of permissible care and approach issues of cruelty and disgust?

Finally, part IV covers the fall. Chapter 8 traces the social and medical processes that, in the early 1880s, led to the demise of lamb blood transfusion.

Hasse and Gesellius were branded as charlatans. Soon they, and later Roussel, disappeared from the medical limelight. But did lamb blood transfusion really fade forever from the medical scene? The Epilogue traces a recurring interest in the 'mighty influence of strange blood' well into the 20th century. We encounter some very special practices. So perhaps Hasse had got it right, after all?

PART I: SETTING THE SCENE

1. Using the blood of others

'No operation in the last two centuries has aroused such high expectations, nor experienced such periods of contempt and oblivion as the transfusion of blood. For more than a century, it virtually disappeared from medical attention and despite being revived fifty years ago, it did not gain ground in a steady march forward but rather followed an ascending and descending curve.'

These words, from 1874, belong to Friedrich Sander, chief physician at the city hospital of Barmen in northern Germany.[1] He was one of many doctors who, in the mid-1870s, shared what a later observer would call a 'widespread [...] fanatical enthusiasm' for blood transfusion.[2] The therapy, Sander noted, had been previously met with both applause and critique, and now seemed to be in vogue again. In hospitals across the continent, hundreds of patients received blood from others and some from the arteries of lamb.

To Sander, the prospect of healing the sick with lamb blood was fascinating. So, too, was the history of blood transfusion. He, and many others, found it important to anchor their trials and tribulations in a dramatic past, and show the foresight and acumen of the pioneers. They traced the origin of transfusion in myths and magic, related the first practical experiences in the 1660s and the ensuing condemnation by medical and church authorities. They then discussed the revival of the therapy in the early 19th century. Many referred to Ovid's' play *Medea* – she withdrew blood from Jason's elderly father Eason's body, infused it with powerful herbs, and returned it to his veins, rejuvenating him. This was not strictly a blood transfusion, nor did Goethe hint at this operation when he let Mephistopheles utter the famous words in *Faust*, 'Blut ist ein ganz besonderes Saft' – another often-used quote. Blood was indeed 'a very special fluid', symbolizing life and death, inclusion and exclusion.

I will follow the example of the enthusiastic doctors and give a historical backdrop to the events detailed in coming chapters. It will help situate the daring experiments with lamb blood transfusion and the acrimonious debates that followed.

The beginnings

The history of actual – not mythical – transfusion starts in the 17th century.[3] The intervention was not thinkable until the theory of blood circulation presented by William Harvey in 1628 had been understood and accepted. One could now imagine that blood introduced into the body's closed system would stay there rather than, as was thought before, be diffused out and destroyed. In principle, too, any artery or vein could function as a convenient entry into the blood stream. Animal experiments now got underway and blood transfusion to humans was the logical next step.

Interestingly, 19th century texts on transfusion sometimes present slightly different stories. Italian authors tend to underline what happened in Italy during the late 17th century. Harvey's work on the continuous circulation of blood had, in fact, been conducted in Padua, and the concept of blood transfusion was readily accepted by many 17th century Italian surgeons. In December 1667, Guglielmo Riva, chief physician to the pope, performed three public demonstrations of transfusion from sheep to very sick patients. At least two of them survived for a few months. He then made some further transfusions from sheep to men and several collaborators conducted animal transfusion experiments. A few years later, in 1680, the physician Francesco Folli published a detailed description of how to perform a human-to-human transfusion, but this was an operation that he himself never tried.[4] These Italian doctors believed that transfusion would bring nourishment and vitality to the body. They considered it more effective than bloodletting to restore the balance of the body's humours, and ideas circulated that the blood of a healthy young donor would induce vigour and strength into an older recipient.[5]

The 19th century German physicians doing historical overviews also often dwelled on the sheep-to-man transfusions performed in the 1680s by the Germans Balthasar Kaufmann and Matthäus Gottfried Purmann. These noted

1. Using the blood of others   19

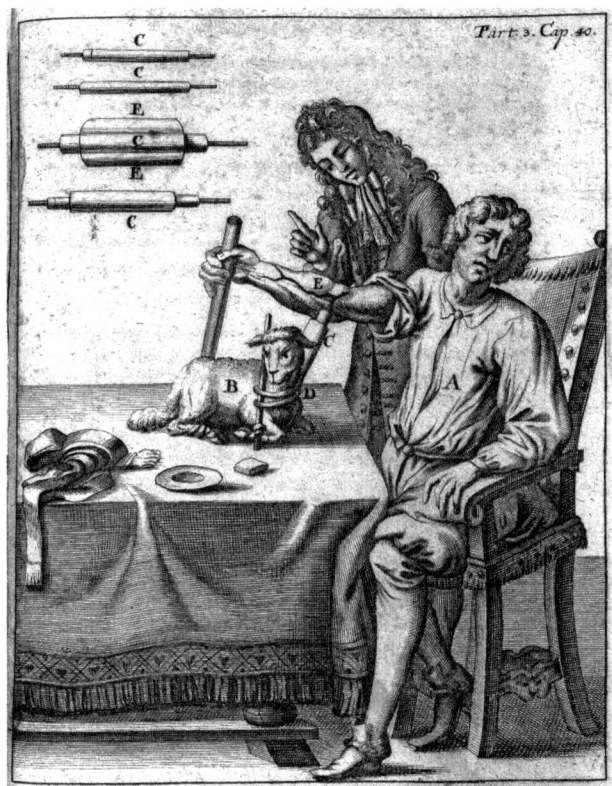

Figure 1. Lower's blood transfusion, 1667. The tubes used to puncture the blood vessels and transfer the blood are at the top left. This illustration is from a 1692 work by the German surgeon Matthäus (Mattias) Gottfried Purmann (Wellcome Collection. CC BY, https://wellcomecollection.org/works/jj7nx24).

no improvement in two scorbutic soldiers but reportedly healed a leaper who, nevertheless, came to suffer from what Purmann called *Schafsmelancholie*, perhaps some sort of sheepish depression.[6]

All 19th century historical overviews, however, gave pride of place to events in France and England that happened somewhat earlier than the Italian and German attempts. In June 1667, the very first transfusion of blood into the veins of a human being took place in Paris. The physician Jean-Baptiste Denis moved blood from a lamb into a young man suffering from anaemia. Some months later, in November 1667, a similar transfusion took place in London under the auspices of the Royal Society. In the presence of doctors and members of Parliament as well as a bishop, Richard Lower and Edmund King transfused blood from a lamb to a man suffering from mental weakness. Thus, the very first transfusions to humans used blood from a lamb and were considered successful. The experiments attracted awe and some ridicule, for instance when Samuel Pepys noted in his diary that they 'did give occasion to many pretty wishes, as of the blood of a Quaker to be let into an Archbishop, and such like'.[7]

More experiments followed in England but in several cases the recipient died. The Royal Society finally saw little value in the procedure. Denis, too, tried some more transfusions, including one with calf's blood to a Swedish nobleman. The Swede was close to the then abdicated Swedish queen Christina. In a letter to her physician she clearly found the idea of a transfusion alluring:

> I think the invention of injecting blood is all very fine, but I should not like to try it myself, for fear that I might turn into a sheep. If I were to experience a metamorphosis, I should prefer to become a female lion so that no one could devour me.[8]

The Swedish nobleman did not make it, however. When another patient died Denis was put to trial but was acquitted. Suspicious colleagues at the Medical Collegium of Paris soon prohibited transfusions, followed by a ban from the Catholic Church. To move blood into humans was to set oneself up as an equal to God with unknown consequences. 'Opponents warned of the risk of transferring the beastly spirit of the donor, which would transform the very nature of man, acquiring the instincts and behaviour of the animal', later historians note.[9]

During the next 150 years, there would be very few attempts to move blood into humans but many animal experiments. These led to advances in the understanding of the components of blood and the role of oxygen in respiration.

One may wonder: How did the physicians writing overviews in the 1860s and -70s know about this early history? Had they read Denis' accounts of his struggles or the *Proceedings of the Royal Society*? No, more likely they had perused the very thorough history of blood transfusion published in German in 1802 by the Danish doctor Paul Scheel, or read the equally detailed follow-up by J. F. Dieffenbach in 1828.[10] They may also have consulted the overviews published in the 1850s and -60s about more recent transfusions.[11] The procedure had, as Sander alluded to above, returned some fifty years earlier to make a certain, though uneven, progress through the hospital wards.

The return

The 19th century revival of blood transfusion was primarily the work of a young doctor in London, James Blundell. He came to think of this remedy after the experience of standing helpless beside a woman bleeding to death from post-partum haemorrhage. His teacher in Edinburgh, John Leacock, had made experiments with transfusion between dogs, so Blundell now proceeded to do some animal-to-animal transfusions himself. Their success encouraged him, in 1818, to make the very first blood transfusion ever to a human using human blood. The patient died but Blundell's later attempts would be more positive. From the mid-1820s onwards, he and others performed several successful human-to-human transfusions. Patients who seemed close to death, mostly women suffering from severe post-partum bleedings, were thus miraculously saved. An 1834 article in *The Lancet* captured the wonder inspired by this recovery: 'Life seemed to be immediately revived as by an electric spark'.[12]

Blundell's daring endeavour – to use the blood of others to bring very sick patients back to life – meant a break with contemporary medical orthodoxy. The prevailing norm was to *bleed* patients rather than to supply them with new blood. He may have been inspired by the romantic notions of contemporary scientists and physicians bringing the nearly-dead back to life.[13] The borderline between life and death was then conceptualized as unclear, shifting and difficult to ascertain. Horror stories were told of people buried alive but rescued in the last instance from the grave, and of drowned and seemingly dead persons awakened by medical men. The step to experimenting with blood, the body's own life-giving substance, was perhaps not difficult to take for a romantically-inclined physician. The very same year, 1818, that Blundell made his first transfusion, Mary Shelley published her book, *Frankenstein: Or The Modern Prometheus*. It built on a similar idea of science giving life to the dead. The scientist, Victor Frankenstein, applied the electrical spark of a lightning bolt to a body whose parts were assembled from local graveyards, and so the live 'monster' was created. The next year, 1819, the first vampire study was published to great public acclaim. It used the same theme, now with the

Figure 2. A transfusion with Blundell's 'Gravitator' 1828/29 (Blundell 1828/29, 321).

vampire surviving through blood harvested from other living beings. Its author, John Polidori, was, like Blundell, an Edinburgh-trained doctor. He was physician to Lord Byron and a friend of the Shelleys' and was possibly present at the famous gathering when Mary Shelley's ideas were first aired.[14]

Thus, transfusion had by the late 1920s been performed by some daring English physicians in cases of severe post-partum haemorrhage. The quite complicated operation may be seen as yet another way for educated male surgeons to wrestle power over childbirth from female midwives.[15] More generally, it was part of a revolution in medical epistemology that started in the 17th century and had been refined from the 18th century onwards. As summarized by later historians, the 'infusion of pharmacological liquors in the veins in general, and transfusion in particular, represented a shift to a new therapeutic concept of care: that of rapid intervention to immediately restore the natural state of the body when traditional long-term therapy has not been effective'.[16]

How to most effectively perform this life-giving intervention was, however, a matter of dispute. There is one very concrete inconvenience with blood: it will rapidly coagulate once outside the donor's body. So, how to avoid introducing life-threatening blood clots into the bloodstream of the recipient? This problem was not easily resolved.

Direct or indirect transfusion?

Blundell's transfusions were of, what he called, the 'mediate' kind. His apparatuses – the Impellor and the Gravitator – were brass implements constructed to gather the blood from the donor and then force it, either mechanically or with the help of gravity, into the patient's body. The idea was to simulate how blood circulates naturally in the body. Blundell's purpose was not primarily to avoid blood clots but to retain what he saw as the 'natural vitality' of the blood. Thus, he thought that rather small quantities of blood would suffice.[17]

Physicians in Great Britain also experimented with techniques of *direct* transfusion. By imitating as closely as possible the heart's natural pumping of blood and thus avoid losing the blood's 'living force', they wanted to move it very quickly from donor to recipient. One influential promoter of this idea was the obstetrician James Hobson Aveling. His transfusion instrument made of rubber tubing and some metal connections had by the 1870s become quite widely used in Anglo-Saxon countries. It was then challenged by the more complex instrument for direct transfusion invented by the Swiss physician, Joseph-Antoine Roussel. We will meet both him and his apparatus quite frequently in coming chapters. For now, we may note that Aveling in 1874 was the first, and ultimately almost the only, doctor to perform a lamb blood transfusion in England.

On the European continent, other transfusion methods were tried out. As early as in 1821, the scientists Dumas and Prévost argued in favour of the *indirect* method. To avoid getting partial or total blockage of the blood stream, one should first bleed the donor of a certain amount of blood. Then, through whipping and filtering the blood to be administered one would remove the fibrin that caused coagulation. Finally, the defibrinated liquid could be introduced into the recipient's vein. Nobody dared to test the method on a human patient until 1847. That time it did not work, but some fifteen years later it had evolved into a rather established procedure and was backed up by clinical experience and physiological research. Most influential were the experiments undertaken by the German-trained Danish professor, Peter L. Panum, whom we also will meet again later in this book.[18] He, and others, bled and transfused large numbers of different animals. They argued for the utility of the indirect method and promoted it as a more reliable operation than the direct variant. But not everyone agreed. Many questioned the medical correctness of whipping and filtering the blood, meaning, they feared, killing its vital, life-giving elements![19]

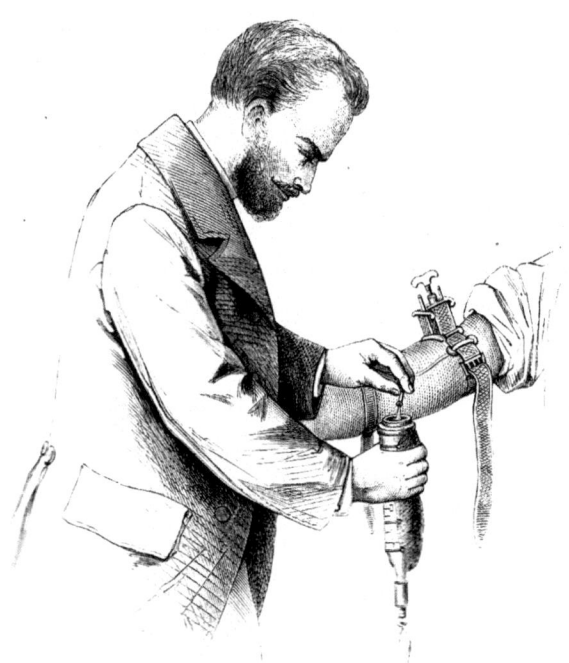

Figure 3. Tapping a donor for blood using the indirect method (Gesellius 1873, 23).

Gaining acceptance

By the early 1870s, enough transfusions, with direct or indirect methods, had been successful for an ever-growing number of physicians wanting to try it. Hundreds of transfusions were made across Europe and several, more or less well-functioning, instruments were devised (including using an English stomach pump, or, in an emergency, a common beer pump, of the kind found in every German village).[20] To some physicians, such as the Belgian doctor, Joseph Casse, this meant that the turbulence of earlier attempts was now a thing of the past: A therapy that had once been 'madly advocated by some, excessively criticized by others, condemned and praised in turn, forgotten for a very long time' was now, he argued, seen as a fairly harmless operation, if properly conducted.[21] The German physician, Heinrich Leisrink, was even more enthusiastic:

> There are not many operations which in such an eminent sense deserve to be called lifesaving as transfusion [...] so simple in its technique, so safe in its execution.
>
> Hit by a sudden, enormous loss of blood, a human being lies on a bed, breathing only laboriously, with wax pale face, and a barely noticeable pulse. Around are relatives frightened to death, expecting the end in any second. Finally, the long-awaited physician arrives and explains, after a short examination, that the patient can be saved by this operation. Everyone volunteers to provide the blood. [Soon] new life runs through the veins of the almost-dead beloved. The face reddens anew, the pulse rises, the central organs are supplied with fresh blood; as if touched with a magic wand, the scene is changed, the person is saved.[22]

Others, however, still saw transfusion as a daring intervention, to be attempted only when no other remedy had worked. But there were problems. How long could you wait until it was *too* late? And if there was no willing donor present, should the doctor offer his own blood – though might he not then himself lose consciousness and control?[23]

Scientific knowledge of the physiology of blood was expanding but still uncertain. Crucially, it would take nearly a half-century before knowledge of the existence of different blood groups would effectively influence transfusion practices.[24] Still, the danger of transfusing incompatible blood was not as great as one may expect. Many 19th century patients got blood from near relatives. Later calculations, based on the prevalence of different blood groups in Western/European populations, show that nearly two-thirds of the mid-19th century transfusions would have passed as compatible.[25] Why some transfusions failed was at the time attributed to air bubbles having entered the blood stream, doctors performing the transfusion too rapidly or with too much blood, or the fact that the patient was on the verge of dying anyway.

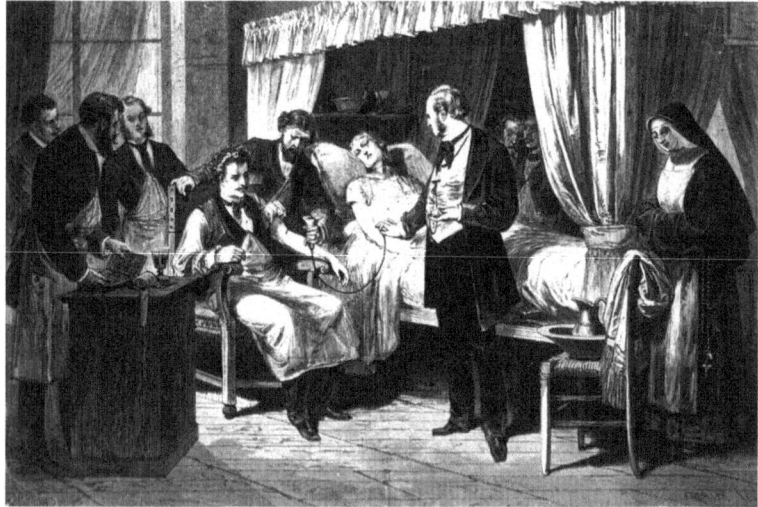

Figure 4. A blood transfusion at the Hôpital de la Pitié, Paris, in 1874. The presence of a nun may indicate that the intervention was no longer prohibited by the Catholic Church (Harpers Weekly, June 4, 1874, 570).

Indications for a transfusion varied. Many physicians, especially in Great Britain, followed Blundell's instruction to transfuse only in cases of acute anaemia, most notably for post-partum haemorrhage and gynaecological afflictions. On the continent, doctors were more audacious. Blood transfusion was tried for conditions such as rabies and cholera, asphyxia, intestinal diseases, carbon-monoxide poisoning, sepsis and leukaemia. Here, too, however, obstetrical and gynaecological problems and cases of acute or prolonged anaemia were the most common indications.[26] This prudence was lauded in 1869 by French physician Charles Marmonnier:

We are far from the time when we claimed to heal everything by transfusing blood: madness, phthisis, cancer, skin diseases, paralysis, fever, without any discrimination, without any solid physiological principles; when we hoped to modify the morale of a deranged individual by injecting him with lamb's blood, to make a pusillanimous man brave by injecting him with the blood of a lion, to restore to an old man the vigour of his youth by injecting him with blood taken from a robust young man. Fortunately, reason and experience soon diminished the exaggerated expectations produced by the enthusiasm generated by the discovery of transfusion.[27]

As we shall see in coming chapters, this verdict would be reversed only a few years later.

Those who in the late 1860s and early 1870s advocated blood transfusion may have fought over what exact method or instrument to use. Still, they all agreed on one thing: *only human blood* could be used for transfusions to humans. Blundell set the tone in the 1820s when he jokingly told his midwifery students why he preferred a human blood donor. In a sick-bed emergency, he said, '[a] dog, it is true, might have come when you whistled, but the animal is small; a calf or sheep might to some have appeared fitter for the purpose; but, then, it had not been taught to walk promptly up the stairs.'[28]

Around the same time, the scientists Dumas and Prévost used animal experiments to show the danger of species-alien blood. Dieffenbach, Magendie, Panum and other physiologists followed suit. Their experimental animals were starved for days, emptied of blood, then transfused with alternately species-similar and species-alien blood. The physiologists injected horses with blood from dogs, transfused sheep blood to ducks, cow blood to cats, bird blood to frogs, and so forth.[29] Their reactions were recorded and the animals, if not already dead, were killed, then dissected, and their urine, liver and blood components studied in detail under the microscope. By late 1860s, the physiologists had established what to them was an indubitable truth: only species-similar blood could be used for transfusion – all else was poison! Science had spoken and the issue was closed.

Or, maybe not?

The revival of lamb blood transfusion

It is the 15th of May 1871 in Wilmington, North Carolina. The local newspaper reports about 'a singular operation to save a man's life – that of the transfusion of blood to his body' having just taken place at the city hospital. The most singular aspect of this event was the identity of the blood donor: a six-month-old lamb.[30]

The patient was a man with a gangrene-infected amputated leg. He was now in a comatose state and rapidly sinking. In the presence of the city's mayor, several other gentlemen and assistants, doctors King and Winants transfused the patient with about six ounces of blood from a lamb's severed artery. He felt much better, got some milk-punch and soon fell into a quiet sleep. He continued to improve for about ten days – but then got rapidly worse. The plan was to transfuse him again but the doctors 'failed to get either a human or an animal in time' and the patient died.[31] The transfusion was nevertheless considered successful, Dr Winants concluded, 'as it was very evident the patient would not have survived through the night if the operation had not been performed'.[32]

The event was covered in US and European media. It was greeted with some amazement but soon forgotten. After all, it was not a lasting success. A year and a half later, another such singular event occurred, this time in Naples, Italy. On the 15th of November 1872, Giuseppe Albini, professor of physiology, transfused a thirty-year-old woman exhausted by severe menorrhagia. The thought of using animal blood was not new to him but this was the first time he tried it. The procedure was reported by, among others, the *Obstetrical Journal of Great Britain and Ireland*:

> A gum elastic tube about half a metre in length was inserted into the artery of a lamb and placed in communication with the vein opened in the lady patient. [Albini renounced the use of a syringe and preferred instead] to use the natural pump, the heart of the animal itself, which with vigorous contractions is able to impel a liberal supply of blood into the arm of the patient.[33]

The patient seemed to improve, but then a new haemorrhage occurred. A second transfusion was performed but this time without much benefit, and the patient died shortly thereafter.

The story about the revival of lamb blood transfusion could have ended there – but the next doctor to seize upon the idea that human lives could be saved by animal blood was more resolute. His energetic promotion of the

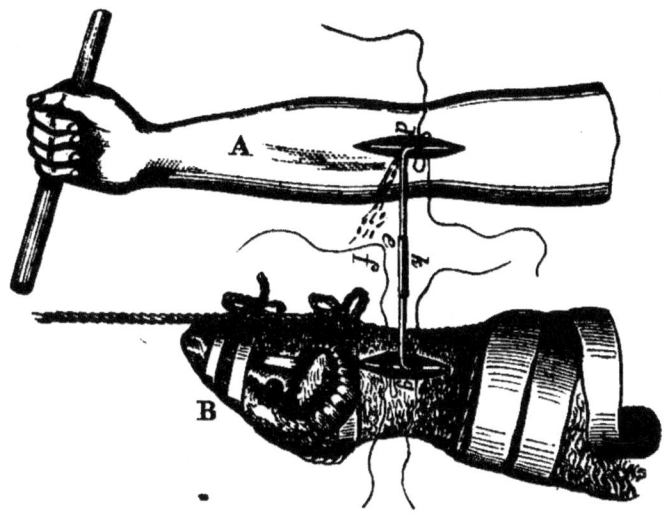

Figure 5. Lamb blood transfusion according to Albini (Albini 1872, 264). Interestingly, he almost exactly reproduces an image published by Paolo Manfredi in 1668.

intervention would, from the autumn of 1873, start an international 'epidemic' of lamb blood transfusion. Was Oscar Hasse aware of the events taken place in North Carolina and Italy? That is not evident.

Hasse's main inspiration was instead a thick volume sent to him by his bookseller who knew of Hasse's transfusion experience with defibrinated human blood. The book, *Die Transfusion des Blutes. Eine historische, kritische und physiologische Studie*, was published in 1873 by Franz Gesellius, a German doctor in St. Petersburg. It was an ambitious, though erratic and polemical, overview of the literature and experience of transfusion since the 17th century. It contained an attack on transfusions with defibrinated blood and ended with a plea for the direct transfusion of blood from the artery of a lamb: it was oxygen-rich, alive, and life-giving! Gesellius' concluding prophecy, *Die Lammblut-Transfusion wird in der Medicin eine neue Aera die – blutspendende – inaugurieren!*, did not fail to make an impression on Hasse. The idea that 'lamb blood transfusion would inaugurate a new era within medicine' was encouraging. And so, on May 26, 1873, Hasse made his first attempt on the young girl, Hermine

Krüger, in Schwenda. Since it was a success, he followed it up with another fourteen lamb blood transfusions, soon to be reported to the world.[34]

The scene was set for the widespread return of a 17th century medical innovation. But who were the main actors behind the 'avalanche' of transfusion that would soon occur? They were many, and not always in agreement. Yet some stand out. I will focus on three central protagonists who personify the experimental, controversial and sometimes successful experience of the 1870s' transfusions. They inspired followers, irritated opponents, and influenced medical practices across Europe and the USA. So, onto the scene I now call Oscar Hasse, Franz Gesellius and Joseph-Antoine Roussel.

2. Ambitions and connections

Innovation, it is sometimes said, is the product of transgression. It entails crossing boundaries, challenging taboos, finding solutions not readily accepted by established hierarchies. It requires ambition and imprudence. The progress of blood transfusion in the 1870s – and its main protagonists – may be seen in this light.

Three medical men would in the early and mid-1870s put their mark on transfusion history: Oscar Hasse, Franz Gesellius and Joseph-Antoine Roussel. Their contributions were admired but also strongly contested: they were complicated, dangerous or simply bizarre. The three men were to a certain extent outsiders, a position they tried to overcome. They were daring and ambitious but otherwise quite different in character.

I will portray these men here and trace their trajectories up to and including the crucial year of 1874. In coming chapters, we will see how their ideas were put into practical (and sometimes not so practical) use at sickbeds across Europe and the USA. Later, I will reveal what happened to them when the transfusion enthusiasm of the 1870s waned.

The sanguine local doctor

Oscar Hasse was a local doctor who from one year to another achieved world fame for his allegedly successful treatment of very sick patients with lamb blood transfusion. He based these claims on his medical experience in and around the town of Nordhausen in central Germany. Dr Hasse, one contemporary wrote, 'was known in distinguished circles as a highly honourable though somewhat sanguine man'.¹ Meaning perhaps, as the dictionary says, 'optimistic or positive, especially in an apparently bad or difficult situation'.² This may have been a useful disposition given the storm that would blow up around him.

Hasse, born in 1837, was the son of a Protestant pastor in Quedlinburg, Harz. He studied medicine in Greifswald and Berlin, where he got his doctorate degree in 1861. He then worked at the protestant Bethanien hospital in Berlin. He learnt surgical techniques, including how to make tracheotomies on young children suffering from diphtheria, a daring operation. In 1864, he moved to Nordhausen to start a private practice in this small but fast-growing industrial town (18,500 inhabitants at the time), not far from Quedlinburg. He soon had to leave, however, to assist at the age of twenty-seven, like so many other young German doctors, in the 1864 war against Denmark. A couple of years later, he joined the medical corps for the second time in the Austro–Prussian war, and in 1870/71 he took part in the Franco–Prussian War, having been promoted to *Stabsarzt*, i.e. captain in the medical corps. Thus, he seems to have made his mark as a military surgeon, receiving military honours for his work.³

In between wars, he fathered several sons and attended to his clinic in Nordhausen with patients both from the town and the nearby countryside. He had professional ambitions, as witnessed by the publication of the results of his and his Berlin colleagues' tracheotomy operations some years earlier, and his articles received a prominent place in a leading medical journal.⁴ Hasse then turned to another daring operation of great current interest – blood transfusion with defibrinated human blood. His first two cases were presented at the 1869 meeting of the Berlin Medical Society. Thereafter, he made another fourteen such transfusions, which made him somewhat of a specialist in this, still experimental, area.⁵

In 1873, Hasse read Gesellius' just published book on transfusion with, as its author claimed, historical evidence from successful transfusions with blood from lamb.⁶ Hasse was impressed. His transfusions with defibrinated blood being only partly successful, he decided, in May 1873, to try this new – actually old – remedy. When he got positive results, he contacted Gesellius who encouraged him to write an account of his first fifteen lamb blood transfusions (of which only one had obviously failed). So he did. He sent the manuscript to Gesellius' publisher in St. Petersburg before going to a conference in Wiesbaden in September 1873. Until then, he later claimed, he had felt

Figure 6. Rautenstrasse, Nordhausen, in the mid-19th century, a street where Hasse lived until 1870. Engraving by Robert Geissler (Stadtarchiv, Nordhausen, StadtA NDH, Best. 9.1.1./ B4 R.01.17).

quite isolated in his endeavours. He had no insightful colleagues to consult about this non-standard intervention. At the conference, however, his presentation was well received, and he got support and encouragement from several well-known physicians.[7]

Hasse's book, *Die Lammblut-Transfusion beim Menschen*, appeared in early 1874 at the St. Petersburg publishing company. Soon thereafter, in April, he presented his results at the German Surgical Society's Third Congress in Berlin. This led to a lively debate, but the participants reached no consensus as to the advantages or drawbacks of the method. This uncertainty did not prevent a growing national and international interest. Hasse received hundreds of letters and inquiries. The popular press got wind of the good news and patients made veritable pilgrimages to Nordhausen in the hope of receiving this wondrous new medication. Doctors across Europe asked for his support, assisted at his transfusions or were inspired by his example to perform lamb blood transfusions themselves.

The enthusiasm would reach quite astonishing heights, as in this review, by a colleague, of his book:

> Hasse's writing shows [...] an originality that we find only among classical writers [...] His patient histories are better propaganda for transfusion to both physicians and laymen than all previous authors' theoretical explications; they give living images. With no intention to do beautiful paintings, the author has portrayed [the transfusions] in such a way that we can see them before us, and the vividly unfolding scenes will encourage us to emulate them and their brilliant success.[8]

By Spring 1874, Hasse had made some 40 lamb blood transfusions.[9]

Despite caustic remarks by some physiologists about Hasse's scientific credentials (more about this in chapter 6), he was respected among colleagues. He was considered a serious and 'honourable' physician, his work was seen as technically adroit, convincing and bold.[10] 'Unselfish and free from personal ambition, simple and undemanding in his outward appearance, forgiving of human weaknesses, always ready to quietly help others', was the epitaph in the *Biographische Lexikon der hervorragenden Ärzte aller Zeiten und Völker* (1931).[11]

The polemicist

Was Franz Gesellius an 'honourable' man? Some contemporaries doubted it. He had, a colleague noted, *'eine eigenartige Persönlichkeit'*, a peculiar personality.[12] His major publication, *Die Transfusion des Blutes. Eine historische, kritische und physiologische Studie* from 1873, was hailed by some as a solid piece of work. Others thought it an incoherent and unpleasant text, with invectives left and right, falsified quotes, faulty and misleading statistics.[13]

Gesellius was three years younger than Hasse, being born in 1840. The son of a physician, he studied medicine in Greifswald (where he was asked to leave after a duelling incident), then in Berlin and Breslau, where he obtained his exam in 1864. He moved to St. Petersburg, which had a sizeable German community. It is not clear whether he practiced there as a doctor. He wrote and lectured on various subjects, including the need for public urinals and the influence of weather on public health but his far-flung ideas were often greeted with ridicule. He became interested in blood transfusion, invented an odd apparatus for the extraction of capillary blood, and then wrote his 1873 history of transfusion leading up to an enthusiastic defence of lamb blood transfusion.[14]

When Hasse, inspired by Gesellius' book, contacted him, he suggested that Hasse should publish his results with his publisher in St. Petersburg. Hasse did not know at the time that it specialised in theatrical publications, a fact that would later earn him some mockery. After difficulties and delays (Gesellius misleadingly changed the title), Hasse's book finally materialized in 1874.[15] So did a small booklet by Gesellius on the same theme and with almost the same title, *Zur Thierblut-Transfusion beim Menschen*. He had by then performed a couple of lamb blood transfusions himself. They were only moderately successful, something that did not prevent him from considering the operation useful, both in civil and, as we shall see in the next chapter, military life.[16]

Hasse and Gesellius had studied at the same universities, albeit at somewhat different times and they apparently never met.[17] By 1875, and in light of the ongoing debate, Hasse came to regard Gesellius as somewhat of a fortune-hunter and swore to have nothing more to do with him.[18] The other two central characters, on the other hand, had both met and competed with each other. Gesellius saw Roussel in action, for example in February 1874, when he was present at one of the Swiss doctor's public demonstrations of his transfusion technique.[19] A couple of weeks later, they both entered a transfusion competition in St. Petersburg.

At stake was which instrument the Russian military authorities should choose for the army, an important decision since it promised both fame and financial rewards. Some twenty contestants showed their different procedures in front of members of the royal family, ministers, ambassadors and medical staff from all major hospitals of St. Petersburg. Gesellius' performance turned out to be both tumultuous and fatal. After numerous difficulties, a patient suffering from phthisis received blood from a sheep for about ninety seconds; he reacted violently and died a few days later. The attendant experts were appalled. Roussel's transfusion, also with blood from a lamb, was on the other hand (he reported himself) greeted with applause.[20]

By then, Gesellius' erstwhile collaborator in transfusion, Oscar Heyfelder, a German physician who was a medical officer in St. Petersburg, had transferred his loyalties to Roussel. Heyfelder had visited Hasse in Nordhausen to learn more about lamb blood transfusion techniques but soon became an ardent supporter of Roussel's device. He assisted Roussel at several public demonstrations of this apparatus. He also used it to make some transfusions

36 Strange Blood

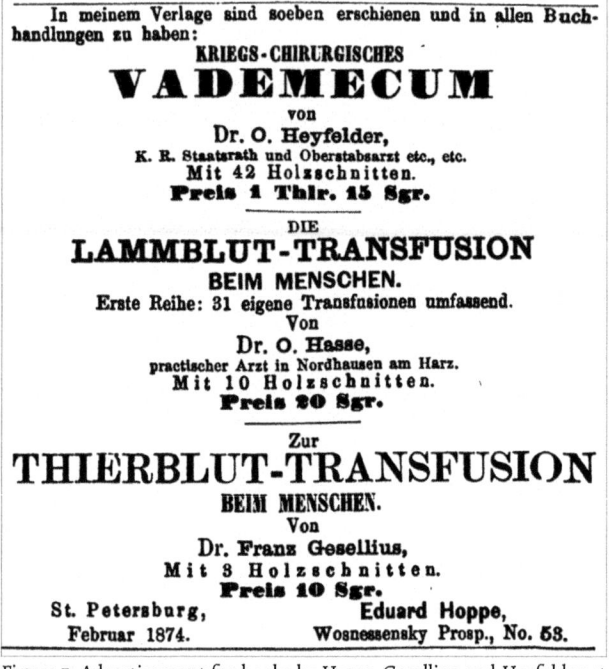

Figure 7. Advertisement for books by Hasse, Gesellius and Heyfelder at the Eduard Hoppe Verlag, St. Petersburg, 1874 (Allgemeine Medicinische Central-Zeitung, 1874, 275). Note that Gesellius' small pamphlet gets the largest text!

himself with human as well as lamb blood. He was, in fact, the officer responsible for recommending the instrument to the Russian military authorities, something that he did in quite celebratory terms.[21]

The entrepreneur

Joseph-Antoine Roussel is credited with being the most ardent advocate of blood transfusion in the late 19th century.[22] Born in Geneva in 1837, he studied medicine in Paris and then travelled the world as a marine surgeon before returning to Switzerland. He practiced in Geneva and established a clinic in the mountains where patients, according to an advertisement, could benefit from 'hygienic and electro-galvanic treatments, cures with grapes and goat-milk, cold baths and hot air'.[23] This interest in non-standard treatments would distinguish him over the years.

Roussel was a quite ingenious inventor of medical instruments. In 1864, at the age of twenty-seven, he designed the transfusion apparatus that would later earn him fame. It had only been tested on animals when Roussel, on a winter's night in December 1865, was hastily called to the home of a young woman. She had miscarried, become unconscious and, it seemed, was rapidly bleeding to death. There was blood everywhere. Present at the bedside were only the girl's sister and a midwife.

Roussel has given several dramatic accounts of what happened then. I have merged them in the passage below.[24]

> No breath, no pulse, no consciousness, but I could hear a faint sound from her heart and decided to try my apparatus. The sister immediately offered to be the blood donor. The various parts of the instrument were assembled, the patient's bloodless vein found with some difficulty, and the donor's vein punctuated. The blood now rushed forward into the dying girl. Nothing happened and I was seized with a terrible anxiety. More pressure on the transfusor's pump and, behold! The girl's heart began to beat somewhat more noticeably. I slapped her face, breast, stomach with a towel immersed in cold water. Finally, the girl's cheeks reddened somewhat, her nostrils widened and suddenly, she took a deep, prolonged and noisy breath. She then had a violent and bloody cough. Still, she did not move. At that very moment, her sister fell brusquely to the ground, fainting more from emotion than from having lost some blood.

The operation was now interrupted, but Roussel calculated that he had moved enough blood, almost 320 grams, into the girl. She regained consciousness, her heart beat more strongly, her lungs moved. She had a stunned look in her yes, as if 'returning from another world'. Soon she smiled, said some words to her sister, and drank half a glass of the warm punch offered by Roussel. Some weeks later she could leave her bed. She later married her lover, and when Roussel saw her she was in excellent health.[25]

Given the particular circumstances of this case, Roussel did not write about it until ten years later and only twenty years later did he reveal why. The situation had arisen, not from a miscarriage but from an abortion, an intervention that in Geneva was punishable by law. The case had been taken

Figure 8. A blood transfusion with Roussel's instrument (Niemeyer 1874, 61).

to court and the midwife was expelled from the country.[26] Roussel had tried to publish the successful experience with his apparatus at the time, but the publication was suppressed, due, Roussel claimed, to the intervention of a rival.[27] Accusations of unfair competition, obstruction and counterfeit would, indeed, follow him and his apparatus throughout his career.

This first dramatic experience caused Roussel to change some aspects of his instrument. In 1867, he presented it at the Exposition Universelle in Paris and, in 1870, just before the war, he demonstrated it to the French War Administration but with no success. He would later mourn the lack of interest since 'the lives of thousands of wounded men might have been saved if the value of transfusion had been fully recognised.'[28] He tried again, unsuccessfully, in 1872, blaming the failure on his lack of influential patrons and connections in France.[29]

Roussel's Austrian colleagues were more interested. In the early 1870s, Roussel spent two years in Vienna perfecting the instrument. He then went on a veritable public relations tour across the continent and made more than a hundred public demonstrations of his apparatus to military and civil authorities as well as to the general public. His transfusions, too, were often public affairs with hundreds of spectators. There, he performed like a life-saving magician 'in the presence of famous doctors, princes, ambassadors, generals, medical candidates and midwives, etc.', one impressed Swedish physician reported.[30] Even the Russian tsar showed an interest and visited patients saved by Roussel's transfusion skills.[31]

Things did not always work out as planned, however. In one unfortunate demonstration in St. Petersburg, and in the presence of an audience of physicians and a visiting English prince with retinue, Roussel's apparatus annoyingly failed. Its valves were not tight enough. A new instrument had to be rapidly fetched from Roussel's hotel. The German physiologist, Leonard Landois, could not refrain from commenting when he got wind of the incident. He ironically predicted that a 'future profound scientist, using all his geniality, may be able to construct an even more complicated instrument [...] with a device to directly measure the amount and speed of the transmitted blood, with an attached thermometer, an electrical self-regulating heater for the blood passing by, a timer to start the whole device, and God knows what more'. Landois' own motto was instead: 'the simpler, the better'.[32]

Roussel's instrument *was* complicated. It included a cupping cup to raise the vein of the donor and a pump to let tepid water into the cupping cup and via a cannula into the receiver's vein. Then there was a lancet to swiftly cut open the vein of the donor, and a tap to let out the water mixed with some blood. With the turn of a stopcock and the help of a balloon pump, the physician could then let the blood flow towards the recipient. This arrangement, Roussel claimed, would prevent the blood from ever encountering air, thus avoiding the problem of blood clots. And the donor would not have his skin and vein cut open.[33]

Despite the complexities, some military authorities were impressed. In January 1874, the Austrian military surgeon Josef Neudörfer suggested Roussel's apparatus for use by the Austrian military authorities. It would, he argued, 'enable the safe transfusion of lamb blood to a large number of patients in only an hour'.[34] Neudörfer thus wanted the apparatus to be used with both human and animal blood. The latter was performed a couple of times by both Neudörfer in Vienna and Heyfelder in St. Petersburg. Roussel, too, twice used

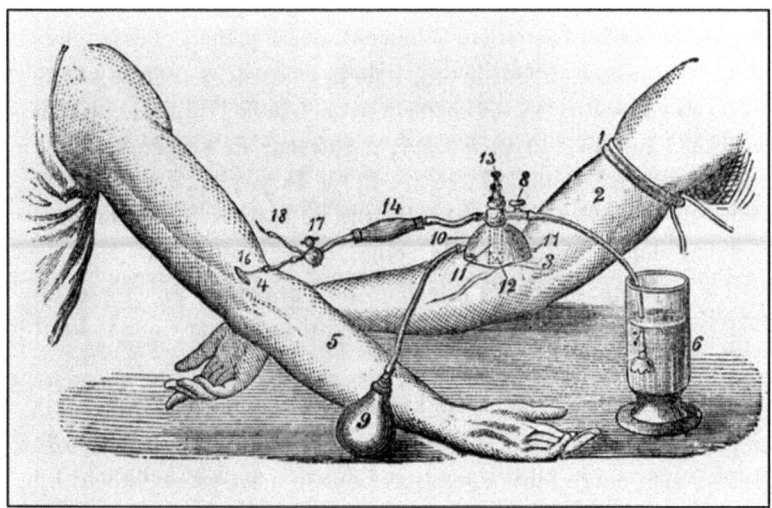

Figure 9. Roussel's *Transfuseur Direct* (Roussel 1877, 44).

a modified version of his apparatus for a lamb blood transfusion. Later, he became more averse to such use, at least with other methods than his own.[35]

In early 1874, after his successful demonstration, Roussel got his instrument recommended for use in the Russian army. This was important, since the Russian authorities would now manufacture the device in such large numbers that Roussel could use them for further demonstrations and sales. Perhaps other armed forces, including the French military, would be interested, at last? I will return to Roussel's efforts in this area.

The context

These three physicians – Hasse, Gesellius and Roussel – were different in character and social position. All three, however, were more or less marginal to the hospital- or university-based establishments of their time. In fact, only Hasse was an integrated member of a national community of physicians. It is unclear whether Gesellius practiced at all as a doctor, and Roussel was a cosmopolitan medical entrepreneur with, it seems, somewhat irregular medical practice. Those following their lead held very different positions within the medical world of the time: they were private practitioners, hospital doctors, asylum psychiatrists, military surgeons... In chapter 6, I will discuss what this diversity meant for how lamb blood transfusion was judged by different medical communities.

To promote their ideas, the three transfusionists employed quite different strategies. Hasse used the channels of professional meetings and journals to demonstrate his surgical acumen and results. He was a modest country doctor, though a fairly established one, given his earlier work in Berlin and his strong military record. Roussel, being a Swiss national, was initially outside professional networks, even in France where he had studied. He therefore turned elsewhere where the interest seemed greater. He exhibited his apparatus at international exhibitions and managed to get both prizes and support from highly placed military surgeons in Austria and Russia, and eventually elsewhere. His tenacity and flair earned him public notoriety, which he used to promote his apparatus. As to Gesellius – he was even more of an outsider to established medical networks. He instead used his connections in St. Petersburg's publishing circles to make his version of transfusion known to the world.

All three were helped by the expansion of both mass and medical media during the mid- and late-19th century. 'English, French and German journals have been teeming with reports of [transfusion] cases, experiments, &c', one observer remarked in 1874.[36] Local and national newspapers, weeklies and magazines geared toward the general public described transfusion experiences in details and with some awe, as did the professional press. The number of medical journals, too, expanded greatly in the mid-19th century to report on meetings at the growing number of regional and national medical societies, as well as on medical reform and on the rapid expansion of medical knowledge. Local doctors had to keep up with clinical and scientific advance. 'Not only was there more science to cover, there were more meetings, more ideas, more politics, and more means', a later historian summarized the situation.[37]

This media – and travel – expansion was made possible by infrastructural investment and innovations: railways, steamships, transatlantic shipping, postal reform. 'Countless magazines, national and international medical congresses, personal contacts of the most varied kind have created a lively intercourse also among surgeons', reported the Austrian military surgeon Theodor Billroth in 1869.[38] Articles were translated, abstracts and reports of experiences in different countries reproduced. Thus, news of new therapies spread surprisingly fast. Italian psychiatrists soon referred to research published in Swedish medical journals and American doctors (especially those with German origin) would imitate what recently had been tried out in Germany. More specifically, Hasse's good results with lamb blood transfusion in the treatment of phthisis and a Viennese physician's use of Roussel's apparatus to cure a mentally ill patient would shortly intrigue and – as we shall see in the following chapters – inspire local doctors far away.

Yet, as Roussel well understood, the real market for transfusion instruments was within the military. The 1860s and early -70s was a period of war. Bloody battles were fought with modern, ever more destructive weapons but under pre-modern medical conditions. The effects were appalling. Limbs were shattered, innards torn, eyes blinded. And blood was shed, gushing violently or trickling slowly towards the soldier's certain death. Military surgeons were desperate. How to save those left dying on the battlefield? Could Roussel's apparatus help? Or would the blood of a lamb carried on the back of a medical orderly bring the seemingly dead soldier back to life? To these visions we will now turn.

PART II: PRACTICES

3. Blood on the battlefield

An unconscious and dishevelled soldier lies on the ground in the arms of another man. A tourniquet is fastened around his left leg. A short tubing with a rubber pump and a dome-like contraption connects the arms of the two men. A uniformed doctor kneels to administer what seems to be a transfusion. There is a satchel on the ground with a Red Cross emblem, and a flask, perhaps containing something to strengthen the men. In the background, two shadowy figures carry a stretcher away from the scene.

The impression is one of a classical Pietà, with the tree behind the men resembling a cross and the wounded man seemingly dead with his hands open towards the sky. The donor holds him in his protective arms; his gaze is one of compassion and care.[1] The doctor's professional gaze, on the other hand, is focused on the arms and the transfusion apparatus. 'God willing will soon new blood and new life flow into the [soldier's] veins', exclaims the author of the article in which this image appeared.[2]

The instrument described in the article about how 'a wounded man is saved by transfusion on the battlefield' was designed by Joseph-Antoine Roussel. He also supplied the image. It appeared in the German family journal *Daheim* (Home) in September 1874. In a later text, Roussel would give detailed instructions for how a transfusion on the battlefield should be carried out, a description that closely resembles the arrangement in the image. He also claimed that his instrument had been used for transfusions at Pontarlier in Eastern France during the Franco-Prussian War.[3] This was not a battlefield but a small town where the retreating French army turned eastwards to enter Switzerland at Verrières. There, in January 1871, more than 89,000 wounded, exhausted and freezing soldiers laid down their weapons and finally got medical attention.[4]

There are, however, no official records of any transfusions with Roussel's apparatus during the Franco-Prussian War. Perhaps we should interpret the

Figure 10. Blood transfusion to save a wounded man on the battlefield (Niemeyer 1874, 60).

enthusiastic *Daheim* article, published some three years after the war, not as a statement of fact but as one of anticipation. It was an advertisement for blood transfusion in general and for Roussel's method in particular, praising its life-saving potential on the battlefield. Roussel himself often argued that if it had been acquired in time by the French (and other) military authorities, much death and agony could have been avoided.

Wars, wars, wars

By any standards, the 1860s was a decade of blood, and the carnage continued into the 1870s.

The list of wars includes the American Civil War, a war in Mexico, the Taiping rebellion in China, and a prolonged war between Paraguay and its neighbours. Europe went through four short but very bloody wars between

1859 and 1871. The Second Italian War of Independence in 1859, also called the Franco-Austrian War, was fought by France and Sardinia against the Austro-Hungarian Empire. Politically, the campaign led to the unification of Italy. Militarily, it was the first war in which both sides used the new technologies of railways, the telegraph and rifled weapons. Medically, it was a disaster with a total lack of trained medical personnel, surgical instruments and ambulances. Wounded soldiers were left lying for days on the battlefield without water, food or medical care. It was after having witnessed the horrors of the battle of Solferino in June 1859 that Henri Dunant began the work that would lead to the International Red Cross.[5]

The Schleswig campaign of 1864, involving Prussia and Austria against Denmark, was short but violent. The next war, the Austro-Prussian War of 1866 with Austria fighting against Prussia and Italy, saw horrendous bloodshed. On the fields of Custoza in Italy, for example, 9,000 Austrian and Italian battle casualties lay unattended for hours. They were difficult to evacuate and no surgeons or ambulances arrived to attend to them.[6] Somewhat earlier, during the American Civil War (1861-65), 620,000 American soldiers died from battlefield injuries and diseases ranging from diarrhoea and measles to typhoid.[7] For every soldier who died of combat-related causes, at least two died of disease.[8]

These wars were waged using the new technologies of capitalism: heavy armour-piercing artillery, machine guns, precision-produced rifles and modern explosives.[9] Such weapons – 'conical bullets animated at a terrible speed by sophisticated rifles, sharp fragments of shrapnel, shells', Roussel observed, 'produce haemorrhage much more often than in the days of old round bullets and massive [cannon] balls'.[10] When limbs were wounded, amputation was not merely the preferred treatment but the necessary one. Soldiers would be removed 'to have their wounds scoured with petroleum and creosote, and their injured limbs sawn off', a later historian notes.[11] Before the widespread use of anaesthesia, the best surgeons could do to improve survival chance was to perform the operation as quickly as possible, thereby minimizing further shock to the victim's system. And surgeons got deft at performing amputations in a matter of minutes. Head wounds, stomach trauma, and infections of the trunk were usually inoperative.[12]

What about blood transfusion? It was discussed as a possible life-saving manoeuvre during the mid-19th century wars, but it was difficult to put into practice. Some military surgeons tried it out while others were more sceptical.

A first attempt was made in 1859 during the Italian War of Independence against Austria.[13] The Austrian surgeon, Ignaz Josef Neudörfer (later a supporter both of lamb blood transfusion and of Roussel's apparatus), was then at the St. Spirito Hospital in Verona attending to injured soldiers with neverending suppurations caused by wounds from rifle fire. The possible dangers of a transfusion – then considered to be an experimental therapy – first held him back. Seeing the patients' desperate condition nevertheless made him dare the attempt. A simple apparatus was improvised, and six wounded soldiers were transfused with defibrinated blood. None of them survived for more than a few weeks. Neudörfer blamed the negative result on the character of the donated blood. It had been taken from a person who was about to have an attack of gout. Therefore, Neudörfer surmised, the blood was full of uric acid and would act as a poison. Still, transfusion had a future, he thought. A French medical journal, commenting on his report, was less enthusiastic: 'A treatment with wine or broth, and stimulating massage […] are emergency measures that are more effective and less dangerous', it claimed.[14] A not uncommon recommendation, as we shall see.

During the American Civil War, blood transfusion was practiced twice, once with success, once not.[15] The American surgeon J. A. Lidell later regretted that it had not been used more. It could have saved those suffering from a state of anaemia and general debility due to great losses of blood, 'from which they could not be raised by even the most nutritious food, alcoholic stimulants, the chloride of iron, or the citrate of iron and quinine'.[16]

In the European wars of the 1860s, some transfusions were made, mostly far away from the battlefield and with no lasting positive effects.[17] How many they were is not clear since most attempts were not published.[18] Among these were the nearly 100 transfusions that Neudörfer allegedly performed during the battle of Sadowa (Königgrätz) in 1866, as it seems with little success.[19] Accordingly, many surgeons were sceptical about the method's feasibility. Making an indirect transfusion with defibrinated blood – the then favoured technique – was considered far too time-consuming in field conditions. And who should supply the blood? Other soldiers were not suitable since their strength was needed for combat. Nor could slightly wounded soldiers be used as a blood reserve. Bloodletting had until recently been a common intervention. Thus, doctors could, in principle, use blood from wounded soldiers' head and breast wounds for transfusion. But by the late 1860s, the norm was to avoid bloodletting entirely or wait until there was an urgent need for it; this method was therefore no longer possible.[20]

War-time modernization

We have now come to the Franco-Prussian war of 1870–71. This was a short, five-month, war that nevertheless led to terrible bloodshed and death. Still, it was the first war in which fewer soldiers died from disease than from enemy fire, that is, on the Prussian side.[21] The German military authorities had learnt from the abysmal conditions of earlier wars and had adopted new ideas about antiseptic techniques, inoculation and anaesthesia. Each soldier was provided with a first-aid kit, including bandaging material and sterile lint as an absorbent.[22] In addition, most German soldiers had been vaccinated and there were only a few cases of smallpox. This should be compared to the 2,000 deaths among the French forces, which were only partially vaccinated.[23]

German military thinking also took inspiration from the changes made to medical organization during the American Civil War. The principle became to evacuate seriously ill and wounded as soon as possible back to Germany, using a coordinated system of railways and hospitals of different capacity, each provided with a different set of expertise – doctors, nurses and medically trained stretcher-bearers.[24]

Did this medical modernization include the use of blood transfusion?

One physician who might have tried transfusions during the 1870/71 war was Oscar Hasse. He had experience both with war conditions and with performing transfusions in civil life. In the 1860s, he had on several occasions left his private practice in Nordhausen to take part, first, in the 1864 war against Denmark, then, a couple of years later, in the Austro–Prussian War, and then again in 1870 in the Franco–Prussian War. By then, the Prussian sanitary corps had an efficient organization and high status. It recruited some of the country's best physicians and surgeons. This was in sharp contrast to other armies, whose medical services in war, as in peace, depended on whoever they could lure into uniform.[25] For ambitious young doctors like Hasse, the war experience meant a chance to learn advanced surgery, something that he got, for example, in the very bloody battle of Königgrätz in 1866. He would there, to use his own words, find an 'overwhelming surgical material'. On the other hand, he later remarked, he had never in the wars of the 1860s felt the need for a transfusion.[26]

It was different in the Franco-Prussian War where he for five months was in charge of a typhus station during the siege of Paris. He had to deal with substantial intestinal bleedings often leading to the patient's death.[27] 'Here,

Figure 11. A battlefield scene with the wounded being tended to and carried away. Lithograph, c.1870 (Wellcome Collection. CC BY, https://wellcomecollection.org/works/fbesqjn8).

the thought of and the desire to make a transfusion gave me many sleepless nights', he remembered. However, the lack of healthy donors was a problem, as was the safety of the transfusion instruments. Those recommended for use in the field were far from ideal. Hasse had, at an early stage, procured a Hüter transfusion box. However, he complained:

> The leather head in the therein included Uterhardt syringe was dripping wet from oil, which also coated the inner side of the syringe's glass point with thick, pearly drops. The smell from the oil was revolting. When I put the syringe in hot water to clean it from the rancid oil that surely would not have had a wholesome effect on the blood to be transfused, the varnish that attached the rubber cap to the glass tube dissolved. The syringe was unusable. Given this situation and the possibly complete dissolution of the closing cap, how easily could particles of varnish or putty have become mixed with the blood and cause the most substantial embolism.[28]

Hence, no transfusions by Hasse during the Franco-Prussian War and very few by other military surgeons. Overall, the German side transfused 19 wounded and 14 sick persons, including some French prisoners. One transfusion took place in a casualty clearing centre, three in field hospitals, one in a permanent war hospital, and the rest in reserve hospitals in Germany. Of the 33 transfused soldiers, 13 were healed, 19 died, and the fate of one patient is unknown.[29]

On the French side, transfusion was performed twice according to official sources. A first transfusion was made in December 1870 on a soldier wounded at Champigny, leaving the soldier dead.[30] A second transfusion, on another soldier wounded at Champigny, took place five years later, in 1875, at a Paris hospital. Here, too, the patient died and, what is more, the donor, a young student 'of delicate constitution' got violently ill and died. This sad experience would inspire, a later article claimed, 'a deep loathing of transfusion' among French military physicians.[31]

Using Roussel's apparatus – or not?

Thus, there were very few transfusions during the Franco-Prussian War. Military surgeons had little training in the procedure; it was seen as uncertain and cumbersome. But many were interested and anticipated its regular use in wars that were certain to come. Here is Dr Bruberger talking to the *Berliner Militärärztlischen Gesellschaft* in 1874 about the need for transfusion apparatuses in war:

> [Imagine] standing at the bed of a soldier wounded for the fatherland, who struggles with death after exhaustive bleeding, and saying to yourself, I could save this man, I could surely have saved him – only this miserable little apparatus is missing. Believe me, I do not pronounce these words carelessly but out of a deep, innermost conviction, when I say: The introduction of transfusion apparatuses for the field hospitals of the army is urgently required.[32]

Bruberger himself favoured an instrument by Schliep, but there was a growing interest in Roussel's apparatus, which the inventor himself considered eminently suitable for war conditions.

This leads us to a rather entertaining polemic. As it turns out, not all doctors were as enthusiastic about Roussel's instrument as he was himself. It demanded much practice to be handled properly. A real indictment against it was delivered by Franz Gesellius. In his 1873 book, he had opposed the use of defibrinated blood, claiming it was 'dead blood'. He also dismissed Roussel's instrument despite its use of direct transfusion with 'living' blood because it was 'very complicated'.[33] A year later, and after having read the *Daheim* article, he was outright caustic about this 'wonderful Rousselian apparatus'. 'What an interesting picture', he said of the battlefield image, and continued:

> Still, I miss in this famous sketch by Roussel a field fire with a kettle filled with water hanging over it. Since where would otherwise the representative of the Red Cross get the absolutely necessary warm water for the Roussel apparatus? Wouldn't a spirit or kerosene fire to boil water be impractical in the open field because of potential wind or rain? Moreover, must not the operator be of an unprecedented dexterity to be able to properly push the apparatus subcutaneously into a bloodless, and therefore invisible, vein?[34]

Gesellius thought it virtually impossible to drag hot water along in war circumstances where often not even cold water was to be found (Roussel, not surprisingly, did not agree). The same objection held for asking soldiers or paramedics to sacrifice their blood:

> [Someone] who does not generously want to give his blood and life in an open and honest struggle for his Kaiser and Fatherland is a scoundrel but he is certainly no scoundrel if he refuses to cut open his veins for friend or foe and, thus, perhaps succumb in phlebitis or get a paralysed arm or, in the best case, become weak and unable to work for days. Just to ask for such an undertaking is, I feel, [...] an inhuman attack on the health and life of one's fellow beings.[35]

So, if Roussel's apparatus was too cumbersome to use and the prospect of bleeding soldiers for direct or indirect transfusion was not realistic, what was there to do? Gesellius had the answer in his 1873 book and, even more explicitly, in his 1874 leaflet: lamb blood transfusion! 'Transfusion with human blood will and must – thanks to animal blood transfusion – in no time belong to the history of medical aberrations!' he exclaimed.[36]

The animal blood alternative

Gesellius was not the first to suggest using animal blood in war. In 1860, the German military surgeon, Friedrich Esmarch, transfused 420 grams of defibrinated calf's blood into a dying soldier. The solider died during the operation despite attempts at artificial respiration.[37] The procedure may also have been tried during the Italian War of Independence (part of the Austro-Prussian War). The Italian doctor, Giuseppe Albini, later reported how he, in 1866, when in Milan and about to leave for the battlefield, decided to bring with him a living lamb in the ambulance. The purpose was to make a transfusion using his 'haemodromometron' instrument. There is no information about whether this actually occurred.[38]

By the early 1870s, however, and as we saw in chapter 1, the question of the usefulness of animal blood for civil or military purposes seemed, once and for all, to have been settled in the negative – a verdict Gesellius was bent on reversing, and so he did. His 1873 book and 1874 pamphlet, together with Hasse's publications and practical experience, changed the debate and inaugurated a period of lamb blood transfusion at sickbeds across the continent. But could it be used on the battlefield? Now ensued a vivid exchange of opinions with many ideas but few empirical examples.

Some military surgeons found the prospect enticing. Several now regretted that they had not thought of using animal blood to revive wounded soldiers during the Franco–Prussian War. Hasse ruefully remembered how 'large herds of sheep grazed unchallenged and unnoticed in the park outside the hospital' where he attended to sick soldiers.[39] Paul Schliep of the Augusta hospital in Berlin recalled how there had been 'columns of sheep accompanying our troops and whose blood would only benefit the enemy's earth'.[40] And Oscar Heyfelder mused in 1875:

> If I, in 1870 and 1871, had known of the curative value and the ease of execution of the direct transfusion of lamb blood, I would have infused new blood into most of the starving and weakened sick from [the battle of] Metz, something that would have made it possible for me to proceed to large operations.[41]

As to what animal to use, suggestions were not limited to sheep. Other animals, too, were present in 19th century battalions: calves, dogs, cows and oxen. The Austrian military surgeon Joseph Friedrich Eckert argued for the convenience of using dog's blood:

> With every troupe and medical service, vertebrates are brought along who have smaller blood corpuscles than humans and who therefore are completely suitable for transfusion. Normally, there is no lack of lamb or calves in an army but if that should be the case, there are always stray dogs around. Certainly, dog blood has the same effect as lamb blood – its blood cells are smaller than human ones. Dogs eat mixed food just as humans do, so would not its blood be more favourable than that of the lamb? Even if there are arguments against it, I am completely convinced that in moments of lethal danger, nobody will be against it and I find arterial transfusions with dog blood on the battlefield to be very appropriate.[42]

Both Eckert and Gesellius advised the operator to cover the donating dog's eyes to prevent its almost human, pitiful and helpless gaze making too deep an impression on the patient. Also the nose should be tied up so that the dog's miserable wailing would not distract those around.[43] Given these complications, a lamb seemed to be a more practical alternative: 'Why not a lamb that, when all is over and done with, may be consumed?'[44]

The task then became how to best organize a lamb blood transfusion under war conditions. For Gesellius it was simple:

> A military medical orderly can accompany each doctor and very comfortably in a leather rucksack carry one or two not-too-heavy lamb, which have already been completely prepared for immediate transfusion.[45]

This suggestion was greeted with some enthusiasm. Eckert, for one, thought it 'a very ingenious idea'.[46] Gesellius gave explicit advice on how the lamb should be prepared, instructions that Eckert embraced and further developed. The animal should be prepared on a board but if none was available, he noted, a rolled coat would do as a support.[47]

The Austrian military surgeon Neudörfer, who had made transfusions with human blood in the 1859 and 1866 wars, also seized upon this new alternative. He constructed a special cannula to be inserted into the carotid of the sheep. This manoeuvre could be made away from the battlefield and hours before it was needed. Once out in the field, the doctor or his assistants could then swiftly and safely transfuse blood from the animal to the wounded soldier. Neudörfer thought that one fully-grown sheep could provide blood for at least four persons. Hence, ten sheep with attached cannulas could be used to transfuse forty haemorrhaging soldiers in a couple of hours. This operation

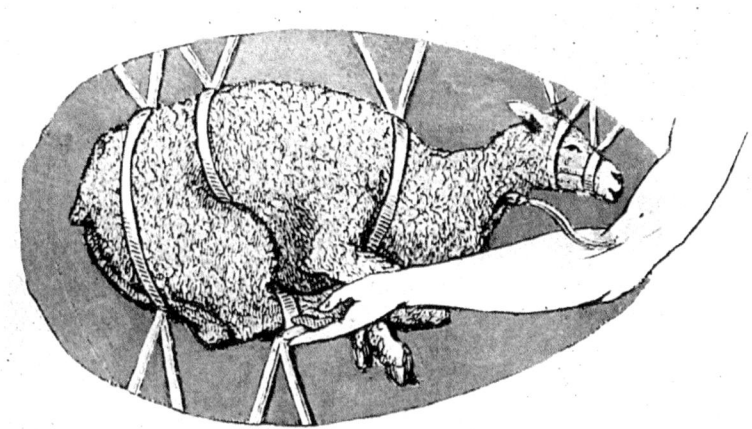

Figure 12. Eckert's suggestion for how the lamb should be positioned for a transfusion in the field (Eckert 1876, 169).

could be performed in a field hospital, but also at a dressing station nearer to the battleground.[48]

Saving the apparently dead

Here, we enter another heated discussion about transfusion in war. *Where* could and should it be done?

Roussel had – with his 'pretty picture' – argued for performing transfusions already on the battlefield, although 'in a sheltered place'.[49] The reason for doing this, he and others argued, was the terrible danger of *Scheintot*, apparent death. Soldiers were declared dead even if they were not. They had lost much blood, were cold, unconscious, had no apparent breathing or heartbeat. But they were alive![50]

The German surgeon von Nussbaum remembered how he, on a cold, dark night after the battle of Orléans in October 1870, struggled to bring soldiers, declared dead, back to life:

> We returned several times, with four or five stretcher carriers, to the wounded who had been left for dead despite that their hearts could still be felt beating and, after having collected, warmed and refreshed them, we brought them back to life.
>
> Loss of blood, exhaustion, hunger, cold and fright were, it seems, the causes of their lethargy. And even if one could not, on the battlefield, use long needles to prick the tip of their hearts, which is the best way to determine death, one could at least [...] put one's ear against their chest, which is very easy to teach any stretcher carrier to do.
>
> For it is too dreadful to think that these poor and brave people would spend an entire night dying in the ditches along the roads while the carriers come and go around them without noticing them. There is no doubt that their stupor will turn into real death if several hours go by without any relief or warmth.[51]

Others were more hesitant. Neudörfer, in 1872, did not think it possible to find and save the apparent dead in the chaos of the battlefield:

> Unfortunately, it is still impossible to distinguish the apparent from the real dead when searching through the battlefield and that will not soon be any better. In battles where there are 2–3,000 dead and 6–10,000 wounded, one human life more or less is hardly relevant. It is not possible [...] to find and carefully determine who is an apparently dead among the thousands of really dead.

Performing a transfusion in such circumstances, he added, was impossible. If, by chance, one would discover such a seemingly dead soldier on the battlefield, he should instead be placed in a horizontal position, made warm and given some drops of rum or wine. 'Of blood transfusion, which would be the most suitable remedy, there is no question in the field.'[52]

To this, Eckert responded: Should we just give up? Should we idly put our hands in our laps and not even try a transfusion because the operation is so arduous to perform and does not always succeed? His answer was: 'Absolutely not – death threatens and there is nothing left to lose', and recommended the use of animal blood in the field.[53] Roussel, too, was in favour of transfusion to the seemingly dead (if not necessarily with animal blood): 'Very often death is only apparent, and even after many hours, it is not to the pit but to the ambulance that one should carry a body that a transfusion might return to life', he argued in 1876.[54]

By then, Neudörfer had changed his mind. In his 1874 recommendations to the Austrian war ministry about what transfusion instrument to acquire (he opted as we have seen in chapter 2 for the one by Roussel), he now thought that a transfusion would do great service during the evacuation of the battlefield. Experience from the war in America, he argued, had shown that not all bodies abandoned immobile on the battlefield were *really* dead. A good part of them were only unconscious due to haemorrhage and in a state of apparent death. They could be called back to life with a transfusion of human or, if necessary, animal, blood.[55]

Again, not everyone was convinced. Even Hasse thought it a bad idea. For the wounded in the dressing station, 'a roast lamb with a good bottle of red wine would be far more appropriate than a living lamb with prepared carotid', he argued.[56] The military surgeon Hermann Fischer, professor of surgery in Breslau, argued that warm, refreshing beverages or champagne, milk, egg, hot water bottles, then repeated ether injections, meat, beer and wine were better therapies in cases of severe losses of blood. And the lamb that Gesellius proposed should be carried on the shoulders of a soldier in every battalion, 'had better stay at home, or should be put in a soup pot in the encampment'.[57] The military surgeon Bruberger agreed: 'Only a total ignorance of the battlefield' could make Gesellius enounce the naive idea that every battalion upon going into action could have a transfusion-ready sheep on the shoulders of a soldier. Instead, one should get the wounded transported away without thinking about introducing foreign blood.[58] And a German reviewer of Gesellius' text had much fun imagining what would happen if the enemy did not show up as predicted and there was no battle – should one then have a new lamb delivered and prepared for the next day? In the future, one could perhaps instead use condensed blood?[59]

Thus, the Gesellius-Eckert-Neudörfer proposal met with both incomprehension and ridicule by other physicians. It was a suggestion made at the writing table, far from the realities of war, and based on an illusion that 'the direct transfusion of lamb blood [could] be the salvation of many human lives'.[60] As it turned out, only one or two of the transfusions performed during the 1870/71 War were made near the battlefield, and they did not use animal blood. The only lamb blood transfusion on a wounded from this war took place far away from the battlefield. It occurred three years after the war and only after Hasse and Gesellius had re-opened the idea of animal blood transfusion. In early 1874, Bruberger and Schliep at the Augusta hospital in

Berlin transfused a severely wounded soldier four times with blood from a lamb but the soldier did not survive.[61]

War and medical innovation

We now leave the wartime world of amputations and apparent death, thundering cannons and the cries of wounded soldiers for the quiet rooms of city hospitals, clinics and spas. Here, lamb blood transfusion was, in the early and mid-1870s, practiced on quite a number of patients. It was sometimes done to counter a disturbing loss of blood but more often as a kind of medicine against such 'civil' conditions as consumption and insanity.

Still, an interesting question remains: how important were the many mid-19th century wars for the development of transfusion, be it with human or animal blood?

The relationship between war and medicine is a paradoxical one. Wars imply horror and suffering, but – some argue – nevertheless bring lasting medical benefit to mankind. Wars provide unique learning experiences for the medical corps. They include opportunities to develop urgently needed new techniques and therapies, they give access to a huge number of wounded soldiers to operate upon and corpses for dissection.[62]

This was partly true in the 19th century, as Hasse noted with respect to his wartime surgical experience. Still, as medical historian Roger Cooter argues, the lessons of war are not necessarily of direct civilian relevance since wartime conditions in many ways are radically different from those in peace.[63] Most 19th century wartime type of injuries, such as large wounds from mortar and shells, were unusual or irrelevant in peacetime. In the chaos of the battlefield, there was often a need for urgency in wound treatment and amputation – but this was much less essential in a well-organized civil hospital with anaesthetized patients, or was impossible to perform in a local doctor's clinic or a patient's home.

As we have seen, transfusions were hardly attempted during the 1860s and 70s wars, for reasons indicated above. Thus, no relevant civilian lessons could be drawn. The impact was rather the other way around: the mid-1870s *civil* experiences of transfusion, including those with animal blood, influenced military thinking. They seemed to promise simple and efficient ways to save lives, even on the battlefield.

Still, one could argue that there was a give-and-take of visions and technical alternatives between the civil and the military spheres. Military surgeons were active in civilian life and promoted new surgical techniques and interventions. And the wars of the 1860s and early 70s had encouraged new medical arrangements and techniques. 'Never before has the military-medical domain gained so much attention, nor has it had such a rich material to take account of', a Swedish surgeon summarized the impact of the Franco-Prussian War. He mainly referred to the improved status and organization of the military medical corps.[64] But much military attention was also, as we have seen above, given to debating and devising transfusion techniques for use in future wars, and in civil life. Military competitions were set up between different methods of transfusion, large military and technical exhibitions were organized and instruments acquired by various military authorities – and the alternative of using the blood of lamb was endorsed by highly placed military surgeons in both Austria, Russia and Germany.

This leads us to the civil experiences of lamb blood transfusion. The intervention was tried out for a variety of indications, most remarkably, for cases of tuberculosis and pellagra. These were terrible diseases that above all affected the poor in society and often led to an early death. How did doctors reason about the need for a transfusion to such patients, how were the transfusions performed? And how did the patients react? Did they get any better?

These questions will be dealt with in the next two chapters.

4. Blood for the lungs

After Hasse's promising results, doctors across Europe and the USA wanted to experiment with lamb blood transfusion on very sick tuberculosis patients. In this chapter we meet physicians, patients and some lamb, learn how to conduct a transfusion, and note its dramatic immediate effects. But was it beneficial in the long run?

It is late August 1874. The season at Mösseberg Spa in southwest Sweden is almost over. Guests are leaving after a summer of invigorating walks in the neighbouring hills and baths in the spa's clear calciferous water. Soon, the birch trees in the park will turn yellow.

Mösseberg's stately main building has just been rebuilt after a devastating fire a couple of years before. The spa mainly attracts a well-to-do clientele but also has a small hospital for poor patients, paid for by donations. In charge is an ambitious young physician, Otto Torstensson. He is eager to try new treatments on the spa's guests, many of whom suffer from lung diseases, most notably tuberculosis, then also called consumption or phthisis pulmonaris. Torstensson recommends inhalation of 'medical vapours', the use of Professor Waldenburg's pneumatic apparatus to enhance respiration, an assortment of thermal baths, electricity therapy and, of course, lots of fresh air.[1]

Recently, news of a possible miracle cure has reached him from the continent: blood transfusion with lamb's blood. Now he wants to try it. A young woman suffering from phthisis is scheduled to undergo the procedure, but she dies on the very morning it was to be performed. Some days later, a new opportunity presents itself.

The farmhand, Carl Jacobsson from Kyrkefalla in Västergötland, 26 years of age, has suffered from general weakness since his childhood. During spring 1874, he became weaker, paled considerably and grew even thinner. He spent three weeks at Medevi Spa without getting any better and upon his return home began to suffer from a persistent cough, had difficulties breathing and was soon so weak that he had to stay in bed all the time. On August 19, he was taken to Mösseberg and found to be afflicted with a complete thickening of the lower lobe of his left lung and peribronchitis of his right lung. He was extremely pale and thin, with no appetite, could not move without support, suffered from fever and night sweat. His pulse was very weak and rapid.[2]

The season is late and there is not enough time for the usual treatments. But the patient's condition, caused by chronic pneumonia or perhaps tuberculosis, is critical and Torstensson decides to try a transfusion. 'I was very keen to make a lamb blood transfusion, for which I had during spring obtained the necessary instruments from Dr HASSE in Nordhausen by Harz', he later reports in a Swedish medical journal.[3] Said and done, a transfusion is performed. It is the very first in Sweden using blood from a lamb and it is deemed a success!

The day after the operation, the patient declares that he feels really well. He eats with a good appetite and soon goes for walks in the fresh air, without any support. After five days, the doctors cannot retain him any longer at Mösseberg and he leaves for home. Five weeks later, Torstensson gets a letter from the patient, where he reports that he is 'so much stronger and has been able to walk several kilometres without any inconvenience; he still, however, has a persistent cough'.[4] When Torstensson meets him in September 1875, a year after the operation, he is 'still healthy and in good shape'.[5]

Encouraged by the good results, Torstensson and colleagues make a second transfusion in November 1874 at the nearby hospital in Falköping, this time on a very sick phthisis patient, a young notary. He seems to recover but then falls into a depression when his mother dies. He leaves the hospital and no more is known about his fate.[6]

Consumption challenges

Phthisis, tuberculosis and consumption – the nomenclature varied and there was no agreement on whether it should be considered one disease or several – was the single most important cause of death in the mid-19th century. Death rates in major cities in Europe and the US were between 800 and 1,000 per 100,000 inhabitants per year.[7] The course of the disease was unpredictable; most victims deteriorated gradually. At its later stages, patients coughed blood, had chest pains, lost weight, became feverish and extremely tired; many would soon succumb. Overcrowded dwellings, malnutrition and lack of care made it a disease of the poor. Many hospitals would not admit them as patients, seeing them as incurable, and many poor families could ill afford to care for the sick when their symptoms became pronounced. It was considered to be a hereditary disease until 1882, when the German scientist Robert Koch demonstrated the existence of the tubercle bacillus. From then on, the disease was seen as contagious, which opened up for preventive measures and the widespread establishment of sanatoria. The death rate would slowly decline.

In 1874, however, physicians like Torstensson, trying to help phthisis patients recover, were years away from such promises. They had an agonising lack of effective remedies; the situation was often without hope. No wonder that they seized upon the news from Germany that lamb blood transfusion was a possible cure for consumption. In early 1874, Oscar Hasse's book had appeared. He there described how he had first transfused five phthisis patients with defibrinated human blood but met with only temporary success. He then made six transfusions with lamb's blood, and now, 'the success was surprising and wonderful. The general health of these patients was soon perfectly satisfactory, and the local symptoms continued to improve steadily'.[8]

The news made quite a stir. Such improvement might otherwise only occur after the patient had spent many winters in a warm, southern climate, away from the cold.[9] This was something many phthisis patients could not afford. A lamb blood transfusion might therefore, as one Swedish observer hoped for, be 'a chance to fight a disease against which we, when it haunts the poorer classes in society, almost always fight in vain'.[10] In a surprisingly short time, doctors across Europe would try out the remedy.

Thus, the good news travelled fast and also reached America, often via resident German doctors. In Addison, Illinois, Dr Hotz expressed enthusiasm but also some doubt:

> A priori, it is true; we could not understand how the transfused blood was to act upon the pulmonary disease. It did not seem very likely that the simple addition to the blood of a few ounces of fresh lamb's blood, would materi-

ally influence so complicated a trophic disorder as the phthisis is the result of. [On the other hand] we are daily using a great many remedies because by experience we know them to be useful although we cannot comprehend their action fully yet. At present time, we do not decide upon the merits of new therapeutics by theoretical speculations but by experiments.[11]

Hotz and colleagues therefore proceeded to experiment with lamb blood transfusion on severely ill phthisis patients (and some others). They transfused a teacher, a butcher, an opium eater cum tobacco dealer, a blacksmith, and several patients of unknown profession, all during the summer of 1874. Before them, in November 1873, Dr Merkel in Boston had tried lamb blood transfusion on a phthisis patient with some success. Dr Sittel in Cincinnati accounted for several, more or less successful cases in the autumn of 1874, Drs. Hoffman and Weyland of Fall River, Massachusetts, told of one successful case in November 1874 and, somewhat later, Dr Briggar of Elvira, Ohio, reported another happy outcome concerning a patient suffering from the last stage of stonecutter's consumption.[12] Half a dozen other cases were reported in US newspapers, some claiming positive immediate results, others ending less well.[13]

But it was in Europe, most notably Germany, that Hasse´s influence would be the greatest. Dr Sander in Barmen transfused a saddler, a tailor and some other workers afflicted with lung disease, though with mixed results – some patients seemed to get better but one died.[14] The German doctor Oscar Heyfelder in St. Petersburg tried the therapy on two women with serious phthisis, of which one recovered. The second patient felt better, but then she 'put aside all rules of carefulness to walk in the midday heat and dust on the Newsky Prospect', got worse and died on June 11.[15] Other doctors who performed lamb blood transfusion on phthisis patients were Flemming in Gadebusch, Schliep and Küster in Berlin, Warfvinge in Stockholm, Brügelmann in Cologne, Thurn in Niederrad, Neudörfer in Vienna, Schmidt in Lahr, Klingelhöffer in Mainz and Molitor in Karlsruhe, to name but a few of those who described their experiences in the medical journals of the time.[16]

They, and others, transfused lamb blood also to patients with other serious afflictions: typhus, anaemia, cancer and leukaemia. Interestingly, the previously dominant usage of blood transfusion – to women suffering from post-partum haemorrhage and gynaecological ailments – was almost entirely absent when it came to lamb blood transfusion. Perhaps there was not enough time to procure and prepare a lamb in the acute case of severe post-partum

4. Blood for the lungs   65

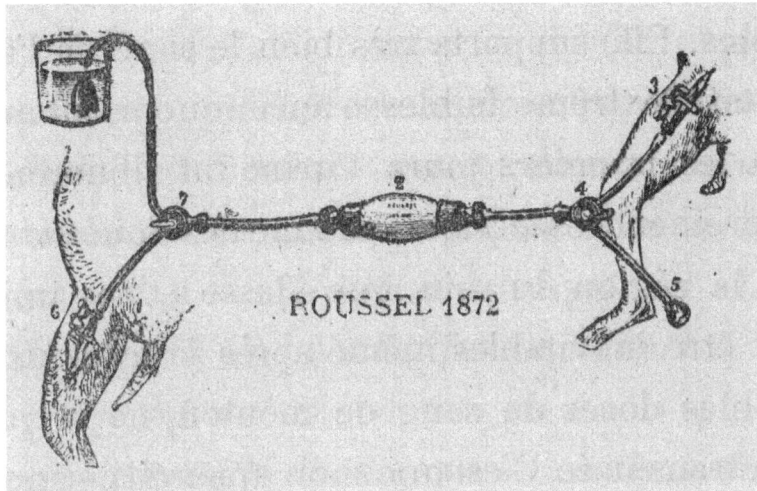

Figure 13. Roussel's modified apparatus used for a vein-to-vein transfusion of lamb blood in the St. Petersburg contest of 1874 (Roussel 1885, 23). It is not clear why the number 1872 is present in the image.

haemorrhage? Or was it seen as dangerous and improper to introduce an animal into the birthing chamber?

By 1874, nevertheless, some British obstetricians, pioneers in the use of transfusion, did consider the use of lamb blood in such circumstances. A couple of years before, the Obstetrical Society of London had formed a special committee to deliberate on the bewildering varieties of transfusion instruments and on the use of whole versus defibrinated blood. Now the committee also discussed animal blood.[17] On April 29, 1874, the well-known obstetrician, James Hobson Aveling, performed the first lamb blood transfusion in England. It was made at the Chelsea Hospital for Women to a woman suffering from a retroverted gravid uterus, but the woman died. Still, Aveling thought that Albini's and Hasse's examples had shown that, when no human blood was available, 'lamb's blood should without hesitation be used'.[18] But there was to be only one more lamb blood transfusion in Great Britain, this time to a haemorrhaging patient at the German hospital at Dalston and from a lamb 'that happened to be on the premises'.[19]

On the continent, on the other hand, transfusion had moved from being a tool used almost exclusively by obstetricians and surgeons to being attempted (now with the blood from lamb) as a general medical therapy. It was used in serious, but not acutely life-threatening, situations, where there seemed to be enough time to procure and prepare a lamb for transfusion. And the favoured indication was phthisis – a choice with a large public resonance.

Hasse's and others' successful transfusions to phthisis patients soon made the popular press. The Mösseberg case in August 1874 immediately caught the attention of a local newspaper. Several Swedish papers also reported about Dr Ziemssen's possible salvation in an Aachen spa of a woman suffering from consumption: 'A young lamb was the innocent animal that was singled out to infuse new strength with its warm blood into this semi-corpse.'[20] In Dresden, there was an onrush of patients demanding transfusion after positive results had been published in the popular press, thereby compelling the hospital's physicians to try the operation.[21] And a probably invented story made the US press in 1874 and even reached newspapers as far apart as Sweden and New Zealand. It told of a case when blood from a goat, for lack of available human or lamb blood, was transfused into a man suffering from consumption. The effect was dramatic – the man started butting the doctor and others present, brayed like a goat, and did not calm down until he was bled and received a second transfusion, this time from an Irishman. His long-term condition was reported to be good, but he shocked his Republican friends by, like most Irish immigrants, becoming a staunch Democrat.[22]

The benefits of lamb blood

Two basic arguments were advanced for why lamb blood could be used in transfusion, one physiological and one practical.

Lamb blood was suitable, it was argued, since its blood cells (called corpuscles) were smaller than those of humans. They could therefore easily travel inside human blood vessels. If no lamb was at hand, Gesellius (and others) argued, one could in situations of real need instead use the faithful servant of humankind – a dog! Cats, on the other hand, were not suitable as blood donors since their corpuscles were too large.[23]

A particular advantage of using lamb was the ease with which a direct, oxygen-rich arterial transfusion could be performed. Arterial blood was supposedly more 'alive' than blood from a vein. But to make a direct arterial transfusion from a human being was a dangerous procedure, something that most doctors definitely did not want to try.[24] A direct transfusion from a lamb's artery was a better alternative, especially if one wanted to do repeat transfusions. As to the sometimes violent reactions arising from a lamb blood transfusion (more about them below), they could be avoided, proponents argued, by a careful administration of the procedure. And did not transfusion with human blood sometimes lead to strong reactions, too?[25]

To this was added a practical argument. Human blood donors were fickle, got frightened or excited when they saw blood streaming out of their bodies. More than one human donor had fainted on him, Neudörfer reported.[26] Getting a lamb, preparing it and performing a lamb blood transfusion was considered a more convenient alternative. Actual practice was messier, as we shall see.

Performing transfusions

I now turn to how physicians described their transfusion experiences at medical meetings and conferences; the patients' versions are, unfortunately, seldom heard. The doctors' reports tell of their patients' desperate condition, their own bewilderment, and their subsequent brave decision to move blood from a lamb to a suffering human being. We learn of their satisfaction when they manage to snatch a patient from the jaws of death and their disappointment at failure despite valiant attempts.

Such case reports were the most important means for clinical physicians to communicate their discoveries to colleagues.[27] The reader (or listener at meetings) was provided with sometimes quite emotional accounts of the patient's suffering, combined with more detached information from percussions and palpations. The reports inform of how the transfusion was carried out and how the patient reacted. They provide pulse and temperature charts and report on the patient's breaths per minute, bowel movements (when, how often, how much) and faeces (is it brown? grey? yellow?), colour and texture of sputum. The amount and colour of urine are reported, sometimes as seen through the microscope to establish the presence of red blood cells and albumen. In only a few cases was the newly invented Malassez method used to count the number of red blood cells before or after the transfusion.[28] Nor was the patient's blood pressure taken since a practical procedure for this purpose had not yet been developed.

The narratives usually begin in a then conventional form. The sick person is situated with regard to social position and gender. His or her name is given, but sometimes only initials, as are age, place of domicile and, often, occupation. At that time, hospitals mainly catered for the poorer segments of society. More well-to-do patients were treated in their homes. Hasse, for example, reported performing transfusions in patients' apartments, in a rectory and at a farm.[29] Many phthisis patients encountered in the case reports were urban workers, but there were also, among others, prostitutes, farmers, a notary, a 'very intelligent school teacher', a bookstore keeper, and officers and soldiers of the Russian, German or Austrian army. The medical histories of parents and siblings are accounted for; many had suffered, and died, from tuberculosis or related diseases.

Having established the need for a transfusion, how then to proceed?

A first necessary step, at least for doctors in private practice, was to get the patient's or the family's permission to transfuse. In most cases, this was an easy task. Hasse reports that patients sometimes begged him for a transfusion or were referred to him explicitly for this purpose. Only in one case, did the family ponder the suggestion for a couple of days. In another, the relatives insisted on a transfusion and the sick person agreed to it only to appease the family.[30]

The next step was to procure a lamb, preferably a young, healthy one of about four to six months. Some physicians found this an easy undertaking, others reported difficulties. Klingelhöffer had a hard time obtaining a lamb and, once he had gotten one, had to be very careful not to bleed it too much, since its owner wanted it back alive.[31] In one case in Sweden, the physician 'tried to find a sheep, but when this was not possible, I obtained blood from two strong fellows'. In another, human blood was resorted to 'since no lamb was at hand'.[32] For Flemming, getting a lamb was no problem in animal-rich Mecklenburg. There were lamb in almost every cottage and even in the small towns, he reported. In larger towns, every butcher could help.[33] But Hasse warned against contacting city butchers – their animals might not be strong enough. Instead, one should cultivate good relations with a competent farmer or shepherd who knew his animals well, had cared for the lamb and its parents, and would handle and transport it in a gentle way.[34] Quite often, the same lamb was used for more than one patient. Mysteriously, the particular breed of lamb was sometimes indicated: an English lamb, a Hungarian lamb, a Merino lamb.

Obtaining the desired *young* lamb was difficult during parts of the year, so some doctors used full-grown sheep instead. This brought an unwanted side effect – the patient would smell oddly after transfusion. The 'intelligent schoolteacher' transfused in Addison, Illinois, was 'haunted' by a strong odour of lamb for at least three days after the transfusion.[35] Neudörfer thought the smell was an effect of what the sheep ate. A lamb being no herbivore but a milk drinker would have less volatile fatty acid compounds in its blood than a full-grown sheep and thus give off less of a smell. For other reasons, too, a young lamb was to be preferred. It was easier to handle, and its blood did not flow as quickly as that of an older animal; this meant less pressure on the patient's heart.[36]

Once these preliminaries had been taken care of, the operation could commence. Present at the bedside were often several assistants and interested medical colleagues. They were useful, Hasse argued, since the various steps of a transfusion should follow swiftly upon one another. The assistants could help open the patient's vein, keep track of time, and steady the lamb during the operation.[37]

We will now follow the detailed account given by the assisting young doctor at Mösseberg, C.H. Björck. It is a story similar to those in most other reports. He first tells of the various implements used. They had perhaps been bought for two thalers from Mechanikus Ockert in Nordhausen, as recommended by Hasse in his book.[38] The apparatus is extremely simple, Björck notes: it consists only of two sets of two and a half inches long glass tubes or cannulas somewhat elongated at one end, two rubber tubes of equal length to fit into the thicker end of the glass tubes, and two brass clips to close the rubber tubes.

Particular attention should be given to the board on which the lamb was to be attached,

> [I]t is of utmost importance that it is bound so that it can breathe freely but not be displaced from its position, and its neck move comfortably close to the patient's elbow joint [...] Thus, you must make the board somewhat broader where the animal's trunk and legs will rest, and narrow it off at the side where the animal's neck and head will rest. Furthermore, you should supply the board with several holes to securely attach the rope with which the sheep is to be tied down.[39]

The next step was to prepare the lamb's artery for the transfusion. Björck describes how this was done at Mösseberg:

Once the lamb had been safely attached to the board, its wool was sheared off near the carotid. A seven cm long incision in the skin uncovered the carotid artery that was then tied up with a ligature. The artery was opened with a small lengthwise cut and the narrow end of one of the glass cannulas was inserted. It had already been attached to the rubber tube and filled with a 1 per cent soda solution. The artery was ligatured to the glass tube and the wound covered with linen patches to prevent coagulation of the blood. The lamb was then left to be carefully guarded by an assistant.

This part of the operation, Björck notes, 'appears to be rather simple [...] however, its practical performance may meet with a number of difficulties'.[40] It is not altogether easy to prepare the lamb's artery, he warns. You risk cutting off the animal's vagus nerve, something that would obstruct its breathing and make it even more restless. Or you might injure the neighbouring *vena jugularis* and cause a major haemorrhage. And even if everything goes fine, it could happen that you, after having made the incision in the carotid, will meet with a rapid torrent of blood that makes the entry of the glass tube difficult. And the pressure of the blood may, even if you have successfully introduced the glass tube and fastened it, push it away again! At every failed attempt blood will be lost. And it is only natural, Björck admits, that the lamb 'makes desperate efforts to liberate itself'.

This, in fact, was what happened at Mösseberg. The successful transfusion to Carl Jacobsson was preceded by a first, failed, attempt a few days earlier. The glass tube was not securely attached, the animal gave a start and blood gushed forward from its artery. The tube was attached again but, once more, slid off the opening in the artery and a stream of blood poured out. At the third attempt, the tube was finally securely fastened but now the animal was so exhausted that it ceased breathing when the transfusion was about to begin, and the operation had to be abandoned. Two days later, it was performed again with a new lamb. This time the transfusion went well.[41]

Given these various difficulties, Björck recommended future transfusionists to first practice on a dead sheep to become well oriented in its anatomy. His advice was taken *ad notam* by another Swedish physician keen to try the therapy: 'The day before the operation I opened up the carotid of a couple of sheep and tried to orient myself in the topography of the animal's neck', he reported.[42] Others made the added precaution of applying a small clamp at the central end (towards the heart) of the lamb's artery to temporarily interrupt the circulation while the artery at the peripheral side (towards the head) was closed by a ligature.[43]

The next step – to prepare the patient's *vena mediana* in the elbow joint – was simpler. For one thing, the vein is just underneath the skin; you therefore have no blood stream to combat, Björck points out. You should just make a small lengthwise incision in the vein, introduce the glass cannula filled with the soda solution into the cut and underbind the vein and the tube with a ligature. This glass tube, too, had a rubber tube attached to its other end, which was closed with a brass clamp. Some doctors reported using local anaesthesia, for example with chloroform, for anxious and sensitive patients. This was probably useful since even trained surgeons would sometimes miss the vein; they then had to try again with often quite painful results.

The transfusion could now begin. Here is Björck again, reporting on how the procedure in Mösseberg was performed: The board with the lamb was brought into the sick room and put on a table close to the patient seated in a comfortable armchair. The clamp was removed from the animal's rubber tube, then a blood stream was allowed to pour out to eliminate possible blood clots. The rubber tube was removed from the patient's glass cannula and the rubber tube from the animal's glass tube was quickly pulled over the patient's cannula. First the soda solution and then the blood from the lamb would pour into the patient.[44] 'This act is, as is easily understood, the most critical moment of the whole operation and demands swiftness and precision', Björck notes.[45]

That the operation was not altogether easy to perform is evident from Torstensson's and Björck's accounts. Others, too, got into trouble. Professor Leube in Jena had studied Hasse's procedure at close hands and thought it simple to perform. Still he got blood clots in the artery cannula and had to change it twice before giving up. Four hours later and with a new lamb, he had problems ligaturing the tube in the animal's artery – it slid hither and thither.[46] Gissler and Wentzel in Pforzheim had to struggle to insert the cannula into the lamb as well as into the patient, and in Boston, Merkel eventually resorted to making an indirect transfusion with defibrinated lamb's blood.[47] Some transfusionists substituted one long glass tube for the rubber tubing and two cannulas, others used a silver cannula for the vein since they feared a glass cannula would break in a sudden movement, still others used a modification of Roussel's apparatus or an instrument invented by the German doctor Schliep that resembled an English stomach pump.

To calm the lamb, Heyfelder took care to wrap a scarf around its head to cover its eyes but leave the mouth free.[48] A US doctor transported the lamb securely enclosed in a sack.[49] Others sedated the animal with chloroform or chloral hydrate, in which case there was no need to strap it to a board. One physician thought it more pleasant for the animal just to have its legs tied together and an assistant keeping it calm by laying his hand on its head. Then, he reported, the lamb behaved with a truly lamblike patience – it was calm and breathed normally during all steps of the operation.[50]

Figure 14. A lamb blood transfusion according to Hasse (Hasse 1874a, inside cover page).

Experiencing lamb blood transfusion

The occasion was not as serene for the transfused patients, however. Their breathing, pulse, temperature and much else were often violently affected. As Dr Roelen in Düren summarized the situation: 'The animal is jolly at once [after the transfusion], but the patient is a piteous sight.'[51]

Here is what happened to Carl Jacobsson at Mösseberg Spa:

After about a minute, his right cheek turned red, his lips blue and the formerly calm patient got very anxious with sweat beads across his brow and difficulties to breathe. He then complained of a terrible backache and a heaviness across the breast. When his symptoms worsened, it seemed best to discontinue the operation. It had by then gone on for slightly more than two minutes.[52]

Hasse recommended to stop the transfusion when the patient complained of difficulties breathing. In the cases reported, this meant after between thirty seconds and three minutes, and most often on the insistent demand of the patient, as in these examples from Sweden:

Towards the end of the transfusion, the patient's cheeks blushed strongly. She was restless and finally screamed out aloud that she could not take it anymore.[53]

The patient soon started to cry about a pain in her lower back, about a pressure over the breast and finally said, 'I think I'll die'. Then the blood stream was discontinued, the cannula removed from the vein, champagne and nerve drops administered.[54]

Accounts from elsewhere tell a similar story of 'extreme agitation, dyspnoea bordering on asphyxia, heightened face and skin colour, bloodshot conjunctiva, cold sweat, strong cyanosis; the patient thinks he will suffocate and makes desperate movements, rises, wants to flee: the countenance is wild, the mouth wide open, the gaze staring, the pupils widened. A violent cough occurs every now and then and is finally alleviated through an expectoration of a bloody froth of mucus. The breathing is by turns rapid, by turns completely absent, it gradually slows down, but a deep, coma-like sleep testifies to the great exhaustion of the organism.'[55]

In some cases, patients had to be sprinkled with water to gain consciousness; wine, port or champagne were given. Most did not want to repeat the ordeal – but it did happen that patients complained about receiving *too little* blood. Hasse reports of the 'blood thirst' expressed by some of his patients:

A patient who finally – and after a long period of wavering hither and thither with fear and excitement – has decided to submit to a lamb blood transfusion and who has [...] endured the often-painful preparation of the vein, such a patient wants something substantial as a reward. Only two, three or at most five tablespoons of blood! That is nothing. You cannot imagine what a blood thirst the suffering patient shows when seeing the beautiful red lamb blood flowing by.[56]

A recurring argument against Hasse's method was the difficulty of ascertaining how much blood had been transfused. Was it only a few drops, or too many? To this critique, Hasse had an answer. After releasing the cannula from the patient's vein, he would let the lamb blood flow for another ten seconds into a measured beaker; he would then multiply the amount thus collected with the time of the transfusion. The critics doubted that this procedure could give an accurate estimate since the increased pressure in the patient's vein would prevent the donated blood from flowing into the patient at the same rate as outside the body and into the beaker.[57] Hasse therefore proceeded to weigh the lamb before and after transfusion, taking care to include any faeces released in the process.[58]

After the initial violent reactions had passed, the patients felt better for a while. Half an hour after the transfusion, they were seized with violent chills that continued for half an hour to a couple of hours. This was followed by profuse perspiration, a high temperature and a rapid pulse and, for some patients, a severe headache that lasted for hours. Thereafter, the patients seemed to feel quite well. They ate and slept, and their temperature became more or less normal. Some had red blood cells or albumen in their urine, others not. Some got itchy urticaria for a few days, others not.

One who experienced this quite painful itching was Dr Redtel, a German doctor who had asked Hasse for a transfusion against his phthisis. 'The itching was intolerable, and I passed very bad nights,' he reported in an English journal. He also experienced terrible pains in the loins that, when they subsided, 'assumed a pulsatile character, synchronous with each arterial beat, so that they alternately increased and decreased, and with each increase of the pain I experienced a sensation as if the blood streamed in with a rush from the femoral vein into the great veins of the abdomen.'[59]

Getting better?

Some doctors describe an almost magical change in symptoms after the transfusion.

Merkel reported: 'Visiting him in the morning, I found him sitting up in bed, just awake from a sound sleep, exclaiming, "It is the best night I have

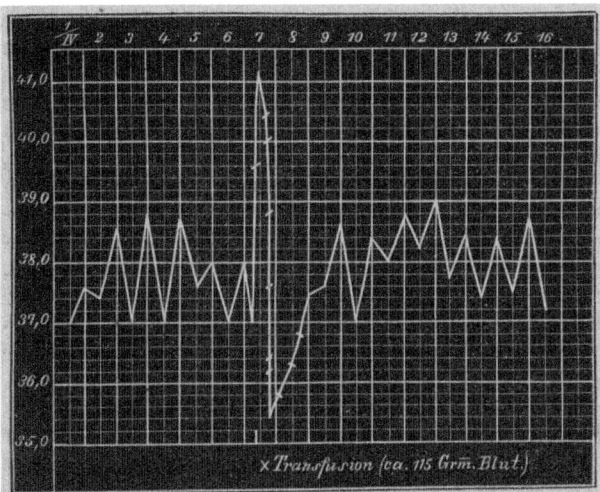

Figure 15. A temperature chart showing patient's increased temperature at the time of transfusion (Fiedler & Birch-Hirschfeld 1874, 556).

had for a year" and asking for something to eat'.[60] Masing, too, told of the sudden great appetite of a patient who had previously hardly eaten at all: at noon a bowl of meat soup, a beef cutlet and large glass of beer, later egg, milk, wine, coffee and tea.[61] Hasse claimed that most of his patients, after transfusion, became very talkative. They joked and made witty remarks; some were so excited that he only with some effort could calm them down.[62] Several patients soon left their beds and took walks outside. One 59-year-old woman, previously a very sick patient, even climbed 'a sizeable mountain' two days after the transfusion. She claimed having experienced no need to rest even once and no problems whatsoever with her breathing.[63]

In some cases, this happy condition lasted. Several physicians reported that their patients had gained weight and gone back to work or other activities. Hasse, for example, told of a 23-year-old tuberculous woman who, some weeks after the transfusion, 'already ha[d] flourished as a lively dancer at two recent harvest festivals'.[64] Von Cube's phthisis patient, transfused by Hasse, went from a very dismal state to almost full recovery. He had gained enough strength from the transfusion to journey to a warmer place when the weather turned cold.[65] For Brügelmann, a combination of better diet, steam inhalation and lamb blood transfusion was beneficial for his patient. He still had an intact left lung; thus, Brügelmann concluded, transfusion seemed to be of value at an early stage of the disease.[66] And Dr Redtel, transfused by Hasse, and the only patient who has given an account of his ordeal, thought the operation useful, although it had not helped him much:

> As regards phthisis it appears that the best results are obtained in those cases where the lung disease was the result of degraded nutrition—e.g., in phthisical women after frequent childbearing, where the greatest and most remarkable results were obtained. Less remarkable, and indeed even doubtful, has been the result in those patients where the disease of the lung was the primary lesion and general health was secondarily affected. This was my case. Yet Hasse assured me even here there have been good results, sometimes weeks or months after the operation.[67]

Other physicians were sceptical. It is well known, one remarked, that phthisis patients, even without a transfusion, suddenly may become strikingly better.[68] This argument was taken up by Herman Alexander Stern, a young doctor who on Christmas Eve 1874 defended a thesis on transfusion. He had earlier that year assisted Hasse with transfusions to three very sick phthisis patients. They initially felt quite well, and Stern asked how this could be explained.

His answer may seem cynical: 'Phthisis sufferers are in some respects the most obedient and grateful of patients.' Any random medicine with a new and different colour was welcomed by them. It gave them hope of a wonder cure with a healing effect greater than all their previous medications. 'And indeed, what a miracle! Maybe an hour after the first spoonful is taken of this new medicine, the patient feels much better and stronger than before'. It should, therefore, he added, come as no surprise that phthisis sufferers, almost without exception, felt so much better after a lamb blood transfusion. They sensed the new, healthy blood seeping through their veins to the ailing lungs, were convinced that blood from such a pious little lamb *must* act as a strong restorative drug:

> The very experience of a transfusion will make such a tremendously imposing impression on the patient that the subjectively felt improvement can partly explain why, after the transfusion, the cough is no longer as agonizing as it used to be, that it does not return so often, that bodily strength has

improved substantially and all this may be a real conviction on the patient's part but it is, I believe, above all created by an unconscious self-delusion.[69]

The argument about the naïve credulity of phthisis patients was a recurring one. There is a tendency among them, the Swedish physician Svensson noted, to improve, no matter what he gave them, as long as it was an interesting novelty. A combination of iron, quinine and arsenic would, for example, have the most wonderful effects, 'and friends of transfusion have not failed to use such restorative means in combination with, or soon after, the transfusion'.[70] Fiedler and Birch-Hirschfeld, too, deplored the delusion among phthisis patients that 'fresh and healthy blood was flowing in their veins'. This was a misconception encouraged by the glowing press reports of this allegedly successful remedy for tuberculosis.[71]

Still, Stern (and others) had to explain the patients' objective improvement, especially the decrease in temperature and the greater appetite. It could perhaps be, Stern assumed, that any increase in fluid in the vessels would lower the amount of fever-inducing substances in the blood. In any case, he added, this effect had been only temporary in the three cases that he observed. During the first few days after the transfusion the patients felt quite well, but within a few weeks all three had died.

A similar fate befell several other transfused phthisis patients. 'Healing was [...] in no case so constant, as some writers have reported and as others who have treated transfusion more theoretically than practically have hoped for and promised', Heyfelder concluded.[72] Also Thurn, initially quite positive, later changed his mind: the transfusion had not fulfilled what his patients had expected from it. They first improved somewhat, but then a rapid deterioration set in.[73]

Still worth trying?

Thus, the experiments to investigate Hasse's claim to have found a cure for phthisis had, in many cases, given a negative long-term result. Still, some patients *did* improve. The Dresden physicians Fiedler and Birch-Hirschfeld were baffled by these results. They had reluctantly made six transfusions just to check whether the therapy was worth trying on lung patients. No improvement occurred, rather the opposite. So they asked: How can we explain the difference between Hasse's successes and our failures?

Fiedler and Birch-Hirschfeld then proceeded to a systematic comparison of cases. The difference between them and Hasse seemed to be due not to how they operated, nor to the amount of blood transfused. The direct effects on the sick person were also the same in almost all respects. Perhaps they had treated patients at different stages of the disease? A careful comparison of disease histories showed no particular benefits to only early stage patients. The difference remained 'unexplained', they concluded, and left it for the future to decide whether Hasse's results would hold. They themselves, however, did not want to make any further experiments since they found it 'inexcusable to henceforth use a procedure for the unhappy phthisikers that is so troublesome and painful for them, a procedure that we from our results cannot give the slightest therapeutic worth'.[74]

Hasse, not surprisingly, did not agree. After his first fifteen cases, he made some fifty more lamb blood transfusions on patients with phthisis and other diseases. The critical debate, the negative results by others and the attacks on his ideas (more of this in chapter 6) caused him to change his arguments for why a lamb blood transfusion might be helpful in some cases of consumption. It would not cure phthisis, he now argued, but it could improve the patient's nutritional status. It could help those phthisis patients who had lost appetite, ceased to eat properly and therefore had degenerated into an anaemic state. A transfusion would give them an appetite and enough strength to respond to other treatments, and recover. Therefore, he now thought that transfusions should be given only to patients in the early, curable, stages of phthisis.

Still, he wondered, what to do with the terminally ill patient, the one who begs the doctor for help and 'clings to what he considers a life-saving transfusion like a drowning man to the last blade of straw!'

> Should we then say to him, 'A transfusion can no longer help you'? You would then have made his last hours of life empty and embittered, when a transfusion could have given him an invigorating ray of hope. For these psychological reasons, you cannot restrict transfusion to only those [in the early stages of the disease].[75]

Should the lamb blood experiments continue, then? This was a not uncommon opinion, even among the sceptics.[76] Perhaps the physiologists were right in their verdict that 'species-alien' blood could not cure consumption, the Swedish doctor Curt Wallis noted in his review of the issue in 1876. Still, he concluded, the issue was 'far from closed' and, above all, it was all too urgent and important to be ignored:

It is here not a question of a new, albeit uncertain, therapy against the common cold or blisters, but humanity's most formidable enemy among diseases. And then, it seems to me that we have no right to leave the question undecided upon because of insufficient investigations but should keep on trying diligently until the issue has been resolved.[77]

5. Asylum experiments

> Threatened by his last sigh
> between blankets, with spasms and in sorrow,
> rests the poor invalid; his errant pupils
> no longer drink the light of the sun!
>
> I approach this languishing patient
> with the piety given to me by God.
> My blood flows into his veins
> and with my blood I give him life.

Thus wrote the Italian professor, Lorenzo Laguzzi, in the local newspaper, *Avvisatore Alessandria*, to poetically depict (from the lamb's point of view!) a remarkable operation that had recently taken place at the city's mental hospital.[1] It was one of many trials in asylums across northern Italy with the blood of lamb to cure the mentally ill. The attempts started in Reggio Emilia in March 1874; hospitals in Alessandria, Imola, Pavia, Pesaro and Brescia soon followed suit. During the next two years, some two hundred lamb blood transfusions were performed in Italy on the insane. Such transfusions to mental patients were something unique; there would be only one or two attempts elsewhere.[2] They were tentative and experimental, criticized but also supported by health authorities and leading psychiatrists.

To me, this was a highly intriguing phenomenon. Why make lamb blood transfusions to mentally ill patients? The idea itself seemed insane. I travelled to northern Italy, visited archives, studied accounts in journals, newspapers and case reports. Gradually a picture emerged; it told of overcrowded asylums and worried psychiatrists.[3] Where to find remedies to soothe their patients' misery, help them recover and return home? The need was particularly urgent in certain parts of Italy. A mysterious disease, pellagra, ravaged the countryside. It was the main cause of insanity in the northern provinces of Lombardy, Veneto and Emilia, covering most of the plains and the Po valley, and claiming thousands of lives each year. Lamb blood transfusion was one of the remedies tested by concerned psychiatrists in the hope that it would help their pellagra patients regain health. More than half of those transfused suffered from this physically and mentally devastating disease.[4]

When I now depict this situation in some detail, I do so for three reasons. The first is to show how lamb blood transfusion functioned in a different medical setting than those I have described earlier in this book. The Italian transfusions almost exclusively took place in asylums. Secondly, they were backed up by institutional forces: lamb blood transfusion was a social, even a political, project. As such, it reflected power relations and power struggles within Italian medical and scientific circles. The disputes even reached the public sphere and the popular press. Thirdly, and relatedly, lamb blood transfusions were here clearly seen as experimental, as something to be evaluated, discussed and perhaps discarded. Thus, the Italian experience opens up, for the next part of the book, the controversy – how could one know if a lamb blood transfusion worked or not, and was it worth the pain and the complications?

To better grasp the particular Italian situation, I will start with the disease itself, the one that the Italian psychiatrists (then often called alienists) hoped could be cured with a lamb blood transfusion.

Pellagrous conditions

In the 19th century, pellagra was an almost unknown affliction outside Italy and parts of southern France. In 1879, up to twelve per cent of the population in Lombardy and Veneto, and slightly less in Emilia were affected by the condition.[5] It was the chief cause of insanity in northern Italy (followed by 'hereditary factors' and 'alcoholism') and it was on the rise.[6] Families and local authorities were unable to cope, and so had to send sufferers to the regional insane asylums. As a result, asylums in areas where pellagra was widely-spread were overwhelmed with cases of 'pellagrous mania'. Still, many were left unattended to at home.[7]

The disease would start innocently enough:

Every year, 'around the time the sun comes into the sign of Aries' [...] the farmer notices a round, dark red, pruriginous spot on the back of his hand that gradually fades and disappears, leaving a patch of gleaming skin. The following year, when the fine weather returns, the patch is larger and the pigmentation darker. These marks then spread to the legs and feet while the hand skin scales off and the small fissures become cracks. The disorder then spreads to the mouth: gums bleed, teeth go black, break, and fall out. The farmer weakens, is taken with nausea, has no appetite. His pulse slackens, head spins, his mind becomes confused. He grows delirious and death ensues.[8]

The first stage of pellagra was skin disease. Physicians adopted the disease's popular label in the Bergamo dialect, *pelle agra*, meaning 'rough skin', after its primary symptom. Terrible headaches and fevers followed. The patient got weaker, sight and hearing were impaired. After the first D – dermatitis – three more stages would follow: diarrhoea, dementia, and, if untreated, death.[9]

Mentally affected pellagrous patients were classified in various ways: they were said to suffer from 'pellagrous frenzy', 'pellagrous melancholia', 'mania due to pellagra' or 'pellagrous monomania', attesting to the inadequacy of the psycho-pathological categories of the time.[10] Some patients would become violent, suffer from delusions, try to chew their tongues off and shout monotonously for hours. Others would be inert with 'no will, no conscience, no word.'[11]

Here is Professor Cesare Lombroso, who would later become the founder of the Italian school of criminal anthropology; he was also a prominent researcher of pellagra. He is quoted in the British *Journal of Mental Science* of 1876. Patients suffering from 'pellagrous insanity', he informs, are easily swayed by their emotions:

A slight insult, the threatening of some trivial danger completely carries them away although they, perhaps, appeared before to be of sound mind. For example, a woman believes herself to be lost because she has missed mass; another person is in despair and goes mad because he has lent a pistol to a friend who will not return it. A woman hears her companions laughing at her dress and becomes insane from grief; another, merely because her husband, a fisherman, is a few minutes late, breaks out into violent mania.[12]

Lombroso had a theory about the origin of the disease: the maize that the peasants lived on was contaminated by a poisonous fungus or mould. This toxin hypothesis was widely accepted by Italian administrative authorities, partly for strategic reasons – in that way the disease would become an ordinary case of food poisoning for which the farmers themselves, and not the state, were responsible.[13] In fact, corn bread or corn-wheat bread, a common food of many poor peasants, was prepared only once a week because many could not afford a daily fire. Huge two-kilogram flat loaves were cooked at high temperature to create a crust, but the inside remained damp and was quickly overtaken by mould and bacteria.[14]

Still, Lombroso was wrong. The main cause of pellagra was structural: it was the inequality, poverty and exploitation in the Italian countryside that caused severe malnutrition, a situation aggravated by damp and insalubrious dwellings. By the 1870s, maize had become the primary crop in six provinces of the newly united Italy. Its increased cultivation brought with it a structural shift in the Italian countryside where large landowners took over most of the land and speculated on what to grow and what to sell. Peasants became labourers working for a (meagre) wage, vegetable plots disappeared, and maize polenta became more than a staple; it became the only food consumed during winter and spring by large sectors of the rural poor. Thus, most sufferers were peasants, day labourers or share croppers. This structural malnutrition hypothesis was indeed suggested by some observers in the mid-19th century but to little effect.[15] That pellagra was caused also by a severe vitamin B3 (niacin) deficiency, caused by the way that maize was prepared for cooking in Italy, would not be convincingly established until 1937.[16]

Pellagra was a disease of the working people. Children normally did not (yet) suffer from pellagra and poor peasants did not live long lives. Pellagra struck women harder than men. For social and biological reasons, linked to the miserable condition of women in the northern Italian countryside, women were more susceptible to the disease than men of the same age. They worked more hours in the field though for less than half the income earned by the men in the family. They did all the household chores, and many had to supplement their income with acting as wet-nurses to more affluent families. Still, they got less to eat than the men, since access to food was strictly hierarchical. The head of the household was served first, then the other working men, and finally the women and children. Thus, many women's dietary intake of the necessary vitamins was inadequate or nil. In addition, the high oestrogen production in women of fertile age induced an even higher risk of pellagra.[17]

In the 19th century, physicians were largely at a loss about how to treat the increasing number of patients. In the early stages of the disease, some curative means were useful. The most common were, according to Dr Brocca in Milan, a meat diet combined with wine, but always in moderate doses given that poor patients were not accustomed to such stimuli. Intestinal flows could be helped by *nux vomica* in increasing doses, sometimes also by arsenic but not by potassium chlorate which some of his colleagues had advocated. Hot baths could help calm the brain of the patients and improve their intelligence, motility and cutaneous sensitivity while cold baths might invoke a terrible terror in the pellagrous insane.[18]

When the patient had reached the third stage of pellagra – dementia, or insanity – most physicians considered the disease incurable.

Figure 16. A corridor in the women's department, San Lazzaro Asylum, Reggio Emilia, in the 19th century. (Courtesy of San Lazzaro Asylum Archive, Reggio Emilia. Album A6 photo n.33 Comparto donne – Galleria Livi.)

Testing transfusion

Then in 1874, news reached Italy of Hasse's good results with lamb blood transfusion for various diseases. So did that of a transfusion (with human blood) in January 1874 conducted by professors Leidesdorf and Neudörfer in Vienna to a severely ill mental patient. It was a success and may well have inspired the Italian alienists.[19]

Here is an account of their very first attempt at lamb blood transfusion on a pellagrous patient: [20]

> Maddalena Selmi, a 44-year old patient, suffering from pellagra, was admitted to the Reggio asylum in the province of Emilia in northern Italy on March 22nd, 1874. She had been sick for over a year and had now reached a state of almost complete decline. She was insomniac, had delusions, spoke nonsense, was maniac. Her skin was yellow, her pulse rapid, she was anaemic and feverish, had no appetite but abundant diarrhoea. The hospital administered treatments and tonics but to no effect. The situation seemed beyond hope.
>
> The young doctor, Augusto Tamburini, then suggested a transfusion to provide nourishment to her organs and help revive her dwindling forces. The asylum director, Professor Carlo Livi, agreed. A first transfusion of 60 grams of lamb's blood took place on April 9th and the patient felt better. A few days later, a second transfusion brought clear improvements. Maddalena now turned lively and gay, regained her appetite and reasoning, could sit up in bed, and showed herself willing and eager for a third operation. This took place on May 3rd and brought an even more significant improvement in her condition. Unfortunately, this was only of short duration, and on May 21st, a fourth transfusion of 60 grams of arterial blood took place. But the symptoms returned, Maddalena got weaker, and on May 25th, the patient died.

Still, the Reggio psychiatrists did not despair; they would try the therapy again. Their next attempts proved more successful and were soon imitated by other asylums in northern Italy.

I see these actions by interested psychiatrists as part of a more general response to medical and societal challenges in Italian society at large. Two events are indicative of their concerns and of the remedies proposed. The first is the First Congress of the Italian Psychiatric Society in September 1874; its published proceedings make it possible to follow the hope and the scepticism expressed about the benefits of lamb blood transfusion. The second is a com-

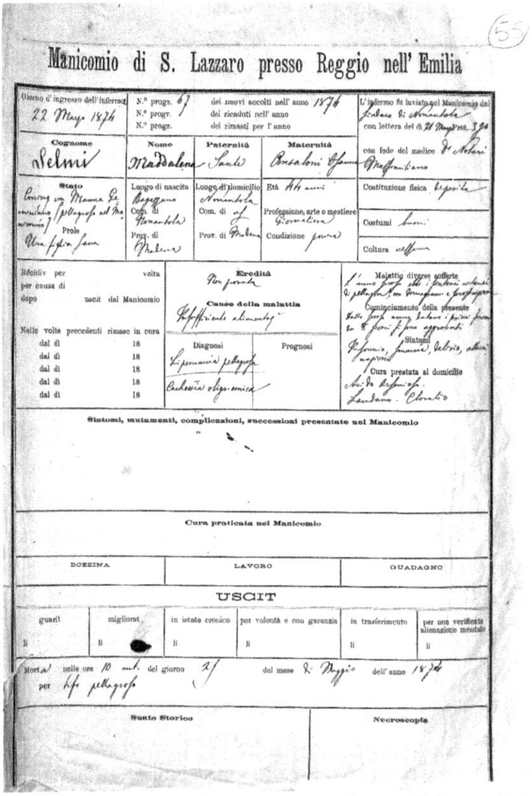
Manicomio di S. Lazzaro presso Reggio nell' Emilia

Giorno d'ingresso dell'infermo 22 Maggio 1874 — N.° progr. 67 dei nuovi accolti nell' anno 1874 — N.° progr. dei ricaduti nell' anno — N.° progr. dei rimasti per l' anno

L'infermo fu inviato nel Manicomio dal Sindaco di Nonantola con lettera del dì 21 Maggio — con fede del medico

Cognome: Selmi — Nome: Maddalena — Paternità: Sante — Maternità: Bersaloni

Stato — Prole — Luogo di nascita — Luogo di domicilio: Nonantola — Età 44 anni — Costituzione fisica — Costumi — Coltura — Professione, arte o mestiere — Condizione povera

Recidiv per volta — per causa di — dopo uscit dal Manicomio — Nelle volte precedenti rimase in cura dal dì 18

Eredità — Causa della malattia — Diagnosi: Lipemania pellagrosa — Prognosi

Malattie diverse sofferte — Cominciamento della presente — Sintomi — Cura prestata al domicilio

Sintomi, mutamenti, complicazioni, successioni presentate nel Manicomio

Cura praticata nel Manicomio

DOZZINA — LAVORO — GUADAGNO

USCIT

guarit li — migliorat li — in istato cronico li — per volontà e con garanzia li — in trasferimento li — per non verificata alienazione mentale li

Morta nelle ore 10 ant. del giorno 21 del mese di Maggio dell' anno 1874 per tifo pellagroso

Sunto Storico — Necroscopia

Figure 17. Front page of patient register for Maddalena Selmi, San Lazzaro Asylum, Reggio Emilia, 1874. (Courtesy of San Lazzaro Asylum Archive, Reggio Emilia.)

petition issued by a renowned scientific academy for studies about the value of blood transfusion in general; its deadline was in February 1875.

I start with the Congress. On its agenda was an assessment of the first attempts at lamb blood transfusion in Italian asylums.

Transfusion and the Risorgimento of Italian science

On the morning of September 24, 1874, a number of prominent Italian psychiatrists gathered at the then newly constructed mental hospital in the small town of Imola near Bologna in northern Italy. They were to witness a transfusion experiment with lamb's blood performed on three emaciated and highly depressed pellagra patients, two men and one woman. The transfusions were made by three psychiatrists using an instrument designed by one of them. They lasted some five minutes each and left the patients momentarily very red in their faces, necks and upper chests. No immediate change in their intellectual functioning was observed by the assembled psychiatrists.[21]

The psychiatrists then reassembled. They were attending the First Congress of La Società Freniatrica Italiana (the Italian Phreniatric Society), which was to discuss the use of animal blood transfusion in cases of severe mental illness. As one of its members phrased it, 'the subject of transfusion is, so to speak, of throbbing current concern'.[22] The mayor of Imola, who inaugurated the conference, was particularly excited about this feature. He saw blood transfusion as 'a daring attempt to return to society many of those unfortunate beings it had rejected' – an urgent and humanitarian task.[23]

There was a sense of excitement about this new endeavour, one of those present reported in the *L'Independente* newspaper: 'The conviction of everyone at the Sunday meeting was that Italy should march proudly because of this new discovery that other nations will applaud.'[24] Italy seemed, after so many years, to again be at the forefront of medical science. Lamb blood transfusion was, thus, no odd or individual initiative. Rather, it should be seen as part of the renaissance of Italian society and culture, and as an expression of the materialistic and anti-religious sentiment of its leading scientists.

Italian resurgence, or *risorgimento*, was the political and social movement that had consolidated the different Italian states into a single nation, the Kingdom of Italy. It started in 1815 and continued through upheavals and wars, such as the 1859 and 1866 wars of liberation against Austria. Unification was completed in 1871 when Rome became the nation's capital. The term Risorgimento also designates the cultural, political and social movement that promoted unification. Thus, the Imola Congress reflected a number of ambitions within some Italian elites at the time: to promote the social and mental health of the new nation, to advance the status of Italian medicine and science, to strengthen the position of scientific psychiatry against superstitions of all kinds.

One aspect of this striving for medical and scientific modernization was a renewed interest in blood transfusion. Italy had, after all, taken an important part in the history of this medical intervention, but one that, Italian scientists lamented, was not sufficiently recognized outside its borders. Was not the Italian doctor, Riva, among the first to do a lamb blood transfusion in the 17th century, yet seldom mentioned beside Denis in France and King in England? Not to forget Andrea Cesalpino who already in the 16th century, according to some, had prefigured Harvey as the discoverer of the circulation of blood, and Michele Rosa who, allegedly, in the late 18th century and before Blundell, had resurrected transfusion.[25] More recently, in 1872, and one year before Hasse and Gesellius, had not professor Albini in Naples re-introduced lamb blood transfusion, while those Germans had got the international credit for it! It was time to reclaim Italy's rightful place in the history, as well as in the present era, of transfusion.[26]

But why use transfusion – and, specifically, lamb blood transfusion – in the treatment of *mental* disease?

The Italian, French and English physicians, who in the 1660s had carried out animal blood transfusions, thought that they would thereby transfer beneficial psychic traits to the recipient. Blood from the docile lamb might calm a violent and mentally deranged patient.[27] Two hundred years later, Italian alienists used different arguments. To them, science, not superstition, should guide the care and cure of the mentally ill. They were positivists; they believed in the power of science to meet pressing social and political needs. Also, they were no political innocents. Several had been involved in the struggle for independence and some would seek political leadership. Now they wanted to use their knowledge of human behaviour, emotions and intelligence to influence social policies, achieve legal and political reform – and wrest the control of madness from religious authorities.[28] They were strong advocates of a humane treatment of the insane: manual work, education and cultural experiences were organized in the asylums to improve the patients' condition. But they also looked for more immediate, medical solutions to the bizarre and desperate condition of the insane. The challenge, as they phrased it, was to revive the mental, moral and social capacities of the mentally ill and help them return to society.[29]

Lamb blood transfusion promised one way ahead. Still, the Italian psychiatrists did not merely look at the clinical results obtained elsewhere. They wanted *scientific* justifications for using lamb blood to cure the insane. Some did animal experiments. Others referred to results by, most notably, the English physiologist, Henry Sutherland. He had shown that mental patients had a large excess of white blood cells at the expense of the red blood cells, and that their red blood cells frequently did not arrange themselves into rouleaux. Such a deterioration of the blood would lead to a very low degree of vitality in mental patients.[30] From this insight, several Italian psychiatrists inferred that a transfusion of fresh blood might have a vitalising effect on the nerves of those who had turned violent or catatonic, dumb or inert.[31]

The choice of a lamb rather than a human donor was, it seems, primarily a practical matter. A lamb was easier to obtain, it was thought to have no transmittable diseases, and it might better support the noise and disorder of a lunatic asylum. In addition, the cutting up of a human donor's vein was seen as an invasive operation that should be avoided. And since the blood cells of a lamb were small enough to pass through the veins of a human recipient and then presumably work just as well as human blood, the choice seemed medically safe.

Figure 18. The Imola Asylum at the end of the 19th century (https://it.wikipedia.org/wiki/File:Cortile_manicomio_ImolaImola.jpg).

First experiences

With these considerations in mind, we re-join the assembled alienists at the Imola Congress. We are to listen to three psychiatrists who have already tried out lamb blood transfusion. Their accounts tell of the patients' agony and of their own bewilderment but also hope for the intervention.

The first case was reported by Professor Carlo Livi from Modena, chief psychiatrist at the asylum in nearby Reggio Emilia and a leading figure within Italian psychiatry. He had been involved in the lamb blood transfusions to Maddalena Sebbi that did not end well. He now could report about some later, more successful, cases. One of them was Andrea Caretti, a thirty-five-year-old man from Modena.[32]

> [He is] an unmarried, timid man, short of understanding who can barely read or write. Four years earlier he had started to work in a billiard hall and although he did not have any bad habits himself, it seems that he, in this infectious atmosphere of smoke and blasphemy, had turned more and more melancholy and morbid; so much so that he often hinted at a desire to kill himself. The death of his mother whom he loved tenderly seems to have increased the sadness of his soul. He locked himself up in his house, became more and more apathetic, dumb and misanthropic, spent hours in bed or crouched in a corner of the house, was filthy, barely eating or talking.
>
> In May [1874], he started to refuse food. Then on one day he tried to stick a spoon in his throat and on another to throw himself from a window. He was taken to the local hospital, and then, on May 30th, to the mental hospital. He was pale, exhausted, extremely thin, unable to stand up, emitted only faint and inarticulate sounds. He had to be fed with a tube. Treatments with cod oil, iron-rich wine, meat etc. had very little effect, and therefore, on June 20th, we resorted to a blood transfusion. He was given a greater dose than usual, about 80 grams of blood, that, however, did not seem to have much effect. Still, on the very same day, he began to eat by himself and with much appetite; he looked less sad and more alert. He started to get up, walk around, talk. His appetite was voracious; his paleness was disappearing.
>
> On July 13, a new and more copious transfusion was made, which lasted twenty seconds. This time, the patient's face became cyanotic, his chest and arms took on a reddish colour, his breath became laboured. He complained of a headache, and of a pain in the back and the stomach. A certain excitement persisted into the evening, but he had no fever. He is afraid of dying but eats with good appetite. From this second transfusion the improvement is even more pronounced, both physically and morally. He answers questions, has no more delusional ideas, nor suicidal tendencies; on the contrary, he

says that he loves life very much, that he desires to return to his family and that he can take care of himself.

Livi was not altogether certain that his patient had been cured by the transfusion:

> Earlier, he was mute, inert, depressed; today he is a man who moves, smiles, works, talks and reasons. He eats but he also eats too much and after the meal he is hungrier than before. He eats dung and grass and stares at the sun. There is something morbid, a darkness underneath, that makes us suspicious.

Despite these question marks, Livi found the result encouraging. Transfusion, he argued, merited further study and experiments.

The next case presented in Imola had been performed in the Alessandria asylum by its chief psychiatrist, G. L. Ponza. It was his endeavours that Laguzzi celebrated in the poem reproduced above.

> Francesco Zunino of Malvicino, a farmer, thirty years old and a father of two healthy children, was admitted to the hospital in Alessandria on June 28, 1873 for pellagra lipemania. He suffered from pellagra-induced diarrhoea; he was thin, sickly, sad, gloomy, silent, without appetite, slept very little, had the tendency to suicide that is almost always present in pellagra patients. All possible remedies had been tried, but in vain.
>
> In desperation, it was decided to perform a blood transfusion. It took place on June 21, 1874, that is, one year after the patient had entered the asylum. It was inspired by the cases in Reggio, was performed by Dr Ponza in the presence of twenty-five other doctors, used the instruments invented by Dr Caselli who was among those present, and done according to the procedure invented by Professor Albini of Naples. As a safety procedure, Dr Ponza first bled the patient of 100 grams of blood. The patient then for twenty seconds received 65 grams of blood from the carotid of a lamb into a vein in his right arm. Two minutes later, he had some trouble breathing, some dry coughs, his face blushed, there were beads of sweat on his forehead, his pulse, which was barely 58 beats before the operation now increased to 75. Ten to twelve minutes later, everything had returned to order and the patient was carried to his room. The very same day he got up and ate. The diarrhoea decreased.
>
> Nine days later, on June 30, another doctor at the asylum, Dr Pacchiotti, performed a second transfusion using the same procedure; the patient received 75 grams of arterial blood into a vein in his left arm. This time he was

not bled in advance. The effect was now stronger: the redness of his face more intense, almost livid, the perspiration more abundant, the breathing a little more troubled; it was feared that he was going to faint. A few minutes later, he returned to his normal state.

Since then, the patient's appetite has reappeared, the diarrhoea has completely stopped, and his forces have returned. Before the transfusion, the patient weighed 61 kilograms; he now weighs 68. The pulse has increased [...] His morale is better, and he is almost cheerful. On July 28, he left the asylum, accompanied by his mother.[33]

Ponza's report included detailed instructions on how to position the patient and the lamb, how to cut open the vein and the artery and how to, with the help of assistants, carry out the transfusion. An accompanying drawing depicts how the patient – Francesco Zunino? – should be seated in a comfortable bed. The lamb, in turn, is less comfortably affixed upside down in a kind of wooden cradle.

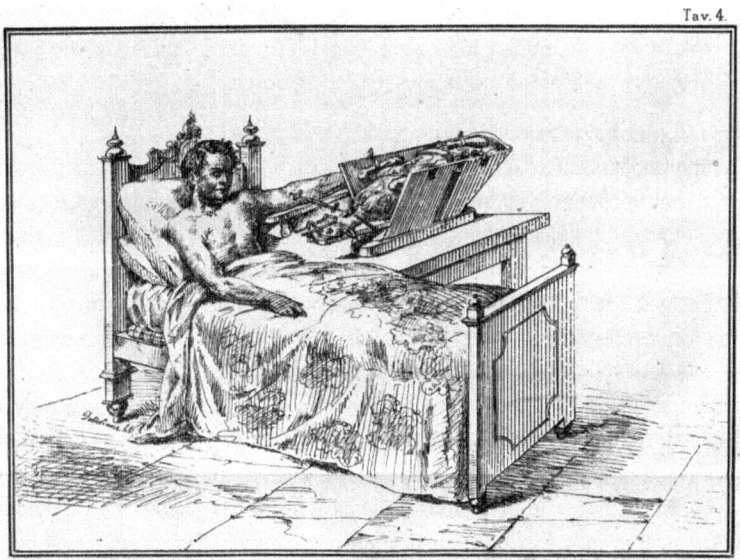

Figure 19. A lamb blood transfusion at the Alessandria asylum (Ponza 1875, between pages 56 and 57).

The third transfusion described at Imola was made by Dr Blessich in Pesaro and reported by his colleague Antonio Michetti.[34] It concerned a forty-six-year-old peasant, Lucia Paglierani. She had been taken to the asylum on May 25, 1874 for pellagra with suicidal delirium. From having been a happy and outward-going person, she had become sad and inward bound, was paranoid and suicidal. Diarrhoea and a lack of appetite had left her emaciated, looking like 'a skeleton covered with thin skin'. On August 12, Dr Blessich transfused 15 grams of arterial lamb blood with no visible side effects. On the following day, the patient said she would like some food, was much less introvert and spoke in a good-natured way. While she had been completely passive during the operation, she now prayed that it should not be repeated. She felt better and wanted to be left in peace. Two weeks later, she left her bed and seemed to be in such good condition that the doctors no longer despaired of her physical and mental recovery.

After these reports and a slightly contentious debate, the assembled alienists agreed on a resolution. It stated that:

> [G]iven that transfusion of blood from the artery of a lamb to the vein of a lunatic patient is neither difficult nor dangerous, and may be performed with ease and certainty, the Congress believes that the efforts of those who have initiated this new form of therapy should be encouraged, provided the treatment is accompanied by a great deal of prudence and, above all, by an attentive study of the indications for and against it.[35]

The resolution was passed unanimously.

The Imola discussion and the reported case histories tell us several things. First, that the patients before the transfusion were seriously ill with delusions and suicidal tendencies. Some had already spent a long time at the asylum but no previous treatment seemed to help. Secondly, and in contrast to how the phthisis patients were treated, the Italian alienists preferred moving quite small amounts of blood into their patients; they also performed several transfusions with some time lapse in between. Thirdly, we learn that, even when the patient seemed to get better, the psychiatrists suspected that this was only a partial or temporary success. Pellagra patients tended to relapse, and an early improvement was often followed by decline. The physicians also realized that part of the recovery might be due to the patients' getting better food and care in the asylum than was possible in their poor homes. Thus, they recognized that lamb blood transfusion, while interesting, was an experimental therapy with uncertain effects.

A transfusion competition

In 1872, the venerable *Istituto Lombardo Accademia di Scienze e Lettere* in Milan issued a competition for the best study of transfusion as a useful medical therapy. At this time only human blood was considered relevant.[36] The competition was initiated by the doyen of Italian medical chemistry, Giovanni Polli, who as early as 1852 had published a transfusion study based on animal experiments. The idea was supported by Professor Andrea Verga in Milan, one of the initiators of the Italian Phreniatric Society.[37] The deadline for the competition was set to early 1875, proving that the issue was of acute concern at the time of the Imola Congress. By then, the possible advantage of animal blood transfusion had also entered the agenda.

The results were announced in August 1875. There were five anonymous contestants. They had made quite different recommendations for how to best perform a transfusion (which the committee called 'a blood graft'): with defibrinated human blood in one case, with animal blood in some others, for mental patients in some proposals, but absolutely not in others. Thus, the contributions reflected the experimental and contested nature of transfusion at the time.

A first prize was not awarded. The prize committee had found faults in most proposals. Instead, a perhaps slightly disappointed committee decided to give three rewards of 500 lire each, 'as an encouragement', to professor Cesare Lombroso of Pavia, Dr Malachia de Cristoforis of Milan, and Drs Rodolfo Rodolfi and Giovanni Battista Manzini of Brescia.[38] Lombroso and de Cristoforis had written lengthy historical overviews with special attention to Italian contributions to 17th and 18th century transfusion history. They also discussed various techniques and indications. De Cristoforis added a report on his six transfusions performed between 1867 and 1873, all with human blood and for both somatic and psychic disorders. Lombroso gave a detailed account of his forty-one transfusions given to eighteen patients at the mental hospital in Pavia between 1869 and 1874. Eight transfusions had been with blood from lamb.[39] These reports, while interesting, will not be discussed here. But the third entry, by Manzini and Rodolfi, is worth a special analysis, since the authors presented their transfusions as a clinical experiment.

Manzini and Rodolfi saw themselves as experimentalists. To them, medical progress depended on experiments and experiment should precede theory.[40] They skipped the lengthy historical exposé, so dear to Italian transfusionists at the time, as well as the discussion of indications and techniques, to focus on their own transfusions made between August 1874 and August 1875. About these, they gave detailed information, first on the choice of patients to transfuse, then on the procedure and the results, and finally they discussed conclusions and recommendations for further experimentation.[41] Thus, their account, though somewhat wordy, is in principle not very different from a latter-day clinical study report.

The Brescia experiment

Giovanni Batista Manzini, born in 1814, was since 1857 chief psychiatrist at the local asylum. He was well-known in Brescia, having received awards for his medical assistance in the 1859 war. He had also acted as psychiatric expert in some highly publicized murder trials.[42]

Manzini's younger colleague, Rodolfo Rodolfi, born in 1827, came from a well-to-do local family; his father was a doctor. Rodolfi himself had, at the early age of twenty-seven, been appointed head of the City hospital. He participated in the wars against Austria, got involved in local politics and was a driving force behind several public health initiatives. Rodolfi was well-known for his dexterity as a surgeon involving some 'innovative and courageous experimentations', as a portrait in a local paper phrased it.[43] These experiments included injecting laudanum, strychnine, alcohol solutions or hydrogen peroxide into animal veins. He also, in one case, injected alcohol subcutaneously into an almost dead cholera patient; the patient first felt better, then died.[44] This experiment led to a conflict with a colleague who claimed that it was without proper scientific value since the cause of cholera was not known.[45]

Rodolfi's and Manzini's experimental venture into lamb blood transfusion was also controversial, something they were well aware of. Given the considerable scepticism among their colleagues, they were quite nervous before conducting their first transfusion:

> We cannot conceal the true trepidation with which we did our first experiments, especially for the one among us [Manzini] who, because of his position as a psychiatrist, had the more direct and serious responsibility.[46]

Interestingly, Manzini and Rodolfi took care to have magistrates witness the transfusion. One may wonder why – to show that the patients were treated well? That no fraud was involved? To give an official stamp of approval? One can only speculate.

The choice of a mental hospital as the site of a transfusion experiment was to them logical for several reasons. First, it had a large enough number of patients – in Brescia about 200 – with symptoms likely to benefit from a transfusion. Patients to be transfused, they argued, should have an illness of long duration and with serious effects on their nervous system, motor and intellectual functioning. A series of treatments should have been in vain. Of the 51 patients chosen for transfusion (that is, a quarter of those at the asylum), more than half had pellagra.[47] They suffered from dementia, hallucinations, suicidal tendencies and some had tuberculosis. Non-pellagra patients were diagnosed with dementia, hysteria, violent mania or alcohol-related insanity. Most patients were highly depressed and intellectually impaired, many were emaciated and more or less depleted by persistent diarrhoea. Blood counts showed them having too few red blood cells or too many white blood cells. This was something that, according to Manzini and Rodolfi, could account for their poor condition but also something that might be improved by a transfusion.[48]

Secondly, a mental hospital had the added advantage of a simple hierarchy. The chief psychiatrist, in this case Manzini, was in charge, and most patients were in no position to protest; they were poor, illiterate and, of course, very sick. With 'so much deficiency of reason', the doctors argued, the director had the responsibility to think and decide for all.[49]

Still, the patient must be willing to participate:

> The operator must persuade the patient of the great utility to be had by a blood transfusion, which is especially important when the patient has not been helped by any other kind of treatment. For our mental patients, words were less effective to achieve their submission and passive assent than were delicacies or some gifts. For this reason, some of them, after the first experiment, spontaneously asked for a repetition of the operation.[50]

Given the fragile condition of the patients, careful clinical preparations had to be made, Manzini and Rodolfi informed. A physician should ascertain that the patient had no circulatory problems or breathing disorders, and no tendency to apoplexy. The patient should be calm and not in convulsions or nervous agitation. Physical preparations also meant that the patient's bowels should have been emptied the same or the previous day. But the stomach should not be completely empty, therefore the patient should receive a light soup, coffee or a broth two hours before the operation. These measures ensured that, in case of vomiting, no unnecessary obstruction would occur and cause distress, and complicate the unfolding of the operation.[51]

During the transfusion, the patient was seated on a chair next to a table where the lamb was positioned. The preferred procedure was to give several transfusions with small amounts (4-40 grams each) since this would lead to a less violent reaction. Still, several patients suffered from cyanosis, fever, involuntary defecation, vomiting or strong chills. Some of them, nevertheless, wanted a repeat of the operation, perhaps to get more treats. Others refused, having suffered 'the onslaught of vomit and the anguish of a threatening asphyxiation'.[52]

All in all, Manzini and Rodolfi made 164 transfusions on forty-nine patients: thirty-two women and seventeen men; two women chosen for transfusion did not get any because their veins were too small. Twelve transfusions were made with human blood. Some patients got both human and lamb blood, and most got several – up to twelve – transfusions. The doctors tried both venous and arterial lamb's blood, coming out in favour of the first for both practical and medical reasons.[53] They devised an instrument of their own, which they claimed was simpler to use than other techniques and did not scare the patients. It had a simple cannula and a pump to help move the blood from the lamb into the patient.[54]

So, was the experiment a success?

Of the forty-nine patients transfused, eighteen were reported cured, six improved, fifteen stationary, five were still under treatment, and five had died.[55] None of the deaths could be attributed to the transfusion, Manzini and Rodolfi argued. Instead, tuberculosis, intestinal troubles and brain lesions were cited as the cause of death.[56]

One of the cured patients was Pasuqua Ransanigo, a thirty-four-year-old peasant woman, deeply melancholic, who insisted on lying on the floor, had to be fed and did not respond to touch nor speech. No other remedies had worked. In September and October 1874, she received three small transfusions from a lamb's vein and was reported healthy also one year later.[57] Another cured patient was Domenica Ruffini, twenty-one years old, who had been taken to the hospital with pellagra. She suffered from suicidal tendencies, looked like a skeleton, had diarrhoea and tuberculosis. She received four transfusions of venous lamb blood of 5 to 8 grams each, after which she started to work and eat. After a second set of transfusions, this time with arterial lamb blood, she felt so well that she could leave the asylum with her parents.[58]

5. Asylum experiments

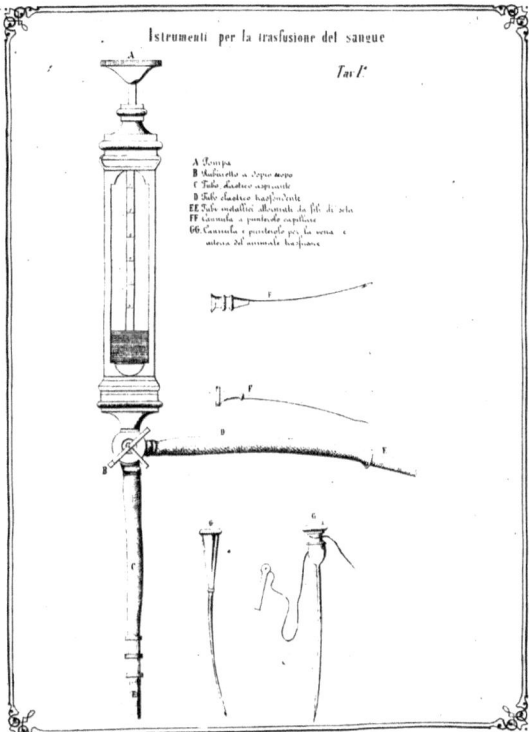

Figure 20. The instrument used in Brescia by Manzini and Rodolfi (Manzini & Rodolfi 1876, 113).

Understanding improvement

How did Manzini and Rodolfi explain such positive outcomes after quite minor transfusions of lamb blood?

Could it be that the very experience of undergoing such an imposing – and even terrifying – procedure had awakened the, until then, drowsy or paralyzed mental faculties of the patients? This 'shock argument' was not uncommon at the time, as analysed later by a medical historian:

In the first few decades of the 19th century, physicians taking a moral approach frequently implemented an additional method to combat diseased modes of thinking. If they believed that a patient could not be rationally convinced of the error of their ways, it was sometimes necessary to shock them into comprehension through a significant emotional experience. These shocks typically took on one of three forms: they could be physical and involve cold showers or some other stimuli; aesthetic and arise from an emotional response stimulated by music or other art; or they could be psychic and involve the staging of an event that resolved a patient's obsession without their knowledge. The latter were enthusiastically undertaken at numerous mental asylums and reported in medical treatises and journals.[59]

Manzini and Rodolfi dismissed this hypothesis. Their patients had, during their often long stays in the asylum, taken part of many, both pleasant and terrifying experiences. They had taken walks in the countryside, had attended music sessions, had witnessed or participated in fights among patients. They had been subject to a series of therapies whose emotional impact, while not equal to that of a transfusion, was quite substantial, for instance, from the suction cups, the scarified cups or the shower. Still, none of this had made them any better while, the doctors underlined, a considerable proportion of those experiencing a transfusion had been cured.[60]

But if it was not the transfusion shock in itself that had cured the patients, what had? Manzini and Rodolfi put forward a *physiological* argument. The transfused blood must have had a stimulating effect on the patients' own blood and thereby on their nervous system and blood circulation.[61] Manzini and Rodolfi were not alone in this suggestion. It was a favourite hypothesis among Italian psychiatrist at the time, rivalling an alternative idea that transfusion was to be seen as a *mechanical* means of adding blood to blood, and thus restoring the patient's blood pressure to a normal level.[62] Both ideas squared well with the prevailing positivist and strongly organicist orientation of Italian psychiatry where, according to Carlo Livi in 1875, 'the so-called mental diseases, those called 'frenopati' or 'frenosi', should be studied only as diseases of the cerebral organ, or of the whole nervous system'.[63] This meant that the transfused blood could be seen as a kind of medicine, like quinine or digitalis, for the nervous system, to be taken in small doses and on a number of occasions. A single transfusion, Manzini and Rodolfi insisted, could not have the desired vital effects.

Given the result of their experiment, what were Manzini's and Rodolfi's recommendations? They concluded their positive report with several interesting caveats. Lamb blood transfusion to mentally ill patients should only be performed, they argued:[64]

- if transfusion was as simple to apply as other remedies, such as laparacentesis, bleeding, electricity, gagging, showering, subcutaneous injections, etc.
- if it did not demand a well-trained surgeon as well as assistants;
- if the transfusion, though not difficult in itself, did not require a myriad of minute and measured attentions and actions, where missing only one at the appropriate time would make the transfusion dangerous, fatal or in vain, so that one had to abort it and start again, either at once or on another day;
- if the apparatus did not have a forbidding effect on the sick patient –even if, as they noted, their mentally ill patients generally had suffered its use with indifference.

Assessing experiments

The Italian lamb blood transfusions to mental patients were controversial. As Manzini and Rodolfi phrased it, there were, on the one side, 'fanatical apostles who endorse [it] as a panacea and a miraculous resource, perhaps without even having tried it', and on the other, 'adversaries that condemn it... [with] derisive sarcasm'.[65]

Let's go back to the declaration from the Imola Congress in September 1874. At first glance, it seems to endorse lamb blood transfusion: it was easy to perform, not dangerous, and those who wanted to do it should be encouraged. But also: there should be a great deal of prudence and, above all, an attentive study of the indications for and against it. This was later interpreted as a 'very reserved and circumspect' decree.[66] Still, psychiatrists in asylums across northern Italy felt encouraged to perform a lamb blood transfusion on a large number of occasions. Others were sceptical. Already at the Congress, some delegates found the procedure too hazardous to be tried out.[67] The debate continued during the next two year, in medical journals, newspapers and the popular press.

Much of the agitation concerned the experience in Alessandria. Here, doctor Ponza and his colleagues had performed about a dozen lamb blood transfusions in the spring of 1874 and some later. Ponza himself was eager to defend his transfusions. He wrote polemical articles and he enrolled illustrious colleagues to witness and to perform transfusions in his asylum. He was supported by the directorate, who paid for a visit to Paris where Ponza did animal experiments together with such luminaries as Malassez and Claude Bernard. Hence, he had enough scientific credentials to gain the confidence of many colleagues but he was also vehemently attacked by other colleagues as well as by the church and the popular press.[68] Some supporters interpreted the polemics against Ponza as a war on scientific progress and a return to dark and obscurantist ages; others saw it as a sign of envy from less prominent colleagues.[69]

Still, similar to the experience of lamb blood transfusion against tuberculosis, those who with some enthusiasm had tried the operation were uncertain, too. Their verdict was contradictory: the intervention was easy but also difficult to perform, it was beneficial but perhaps not so in its effects.[70] Manzini and Rodolfi, being those with the most extensive experience of moving blood from lamb to mentally sick patients, were, as seen above, circumspect despite their positive results. But, just as in the case of lamb blood transfusion against phthisis, they, and others, hoped that further experiments – made 'in the spirit of the new times'– would lead to a breakthrough in the treatment of psychiatric disorders. So far, their trials had shown that lamb blood transfusion *could* be beneficial, if done 'with prudence'.

As professor Carlo Livi, the pioneer of lamb blood transfusion to the insane, expressed it:

> Mind you, we do not believe that we have discovered the wonderful secret of healing such serious forms of frenopathy as pellagra and stupid lipemania. We are used to seeing illusions and hallucinations all day, and therefore know how to guard against introducing them into our own practice. In science we belong to the ranks of sceptics rather than to those of faithful followers and believers. We intend only to try, to experiment. Being certain that science cannot progress or benefit humanity if we do not follow the simple but true canon of that great legislator of human knowledge, Galileo, that is, to try and to try again.[71]

PART III: CONTROVERSY

6. Proofs and refutations

It was not self-evident how the results of lamb blood transfusions should be assessed. Physiologists making animal experiments thought species-alien blood was poison. Practicing doctors were not so sure. They distrusted laboratory evidence when their clinical experience told them otherwise. Neither mode of production of medical knowledge could give definite proof one way or another about lamb blood transfusion. The result was quarrels and confusion.

It is late January 1875. In the Physiological Institute of the University of Greifswald in northern Germany, Professor Leonard Landois is busy completing a series of animal experiments. They were the last of many experiments that he would publish in a large monograph later that year.[1] Landois had, since the mid-1860s, made more than 300 experiments moving blood between animals of different species. He had, for example, injected frogs with blood from dogs and pigs, and transfused dogs and rabbits with human blood and with blood from sheep, guinea pigs, calves and cats.

What emerged from these, no doubt often messy, experiments was that species-alien blood dissolved in the blood of the receiving animal. This then acquired a deep ruby-red colour from the haemoglobin, set free from the red blood cells. Landois saw the dissolution as a clear proof of the uselessness and danger of transfusing species-alien blood.

He also performed what he called pre-transfusion experiments. He used the microscope to check what happened when serum from one animal was mixed with blood cells from another. He found that the globules first adhered together and became spherical. They then lost their colouring matter. Soon only a sticky clump of fibres remained, formed by the red blood cells. This reaction, Landois thought, was due to 'a strange, to us still unknown effect of the mixture with the [serum's] constituent elements.'[2] Twenty-five years later, Landsteiner would find the same reaction when he mixed blood and serum from individuals of the *same* species; he concluded that their blood belonged to different and incompatible 'blood groups'.[3] Of this, Landois, of course, knew nothing. His hypothesis in the 1870s was that such clogging appeared if the blood and serum came from two *different* species. He saw it as yet another indication that transfusion with species-alien blood was extremely dangerous: it would lead to embolism, inflammatory phenomena and, ultimately, death.

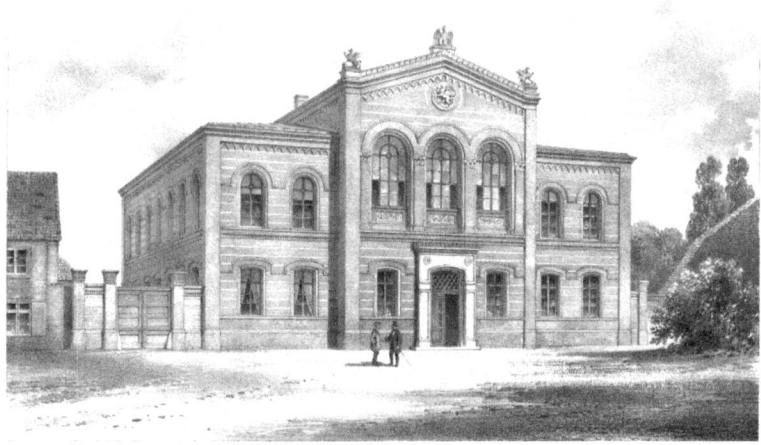

Figure 21. The Anatomical Institute of the University of Greifswald in 1855 (Zeitschrift für Bauwesen 1861, 53). Landois' Institute of Physiology was housed in this building until it got its own in 1888.

Landois' condemnation of lamb blood was echoed by other physiologists. In late 1874, professor Ponfick of Rostock and, in 1875, the Norwegian physiologist Worm-Müller published lengthy critiques based on their animal research; Ponfick had also made one (unsuccessful) lamb blood transfusion.[4] In

1875, too, the Copenhagen physiologist P. L. Panum published a more than ninety-page diatribe against Hasse and Gesellius; he followed it up in 1876 with a further attack when Hasse had responded to his first text.[5]

Thus, the use of animal blood was contentious. A French physician, Louis Jullien, summarized the situation in 1875:

> As we can see, the discussion is vividly engaged. Will Gesellius and Nordhausen [i.e. Hasse] succumb under the weight of attacks coming from so high up? Nobody can currently predict the outcome of this struggle. Let us note here, however, that while the transfusionists rely on observation and put together the most persuasive facts to convince us, the opponents, disdainful of the sick and confined in the heights of physiology, do not put forward a single clinical argument; so that if we had to summarize the state of minds concerning animal transfusion, we would be inclined to write: the clinicians accept it and welcome it; the physiologists condemn it.[6]

At stake in these disputes was *what kind of evidence* should determine the future of transfusion as a medical therapy. For physiologists and their supporters, animal trials had clearly shown that lamb blood transfusion had little foundation in science; it was a dangerous experiment on fragile patients. On the other side, '[n]o clinical practitioner would let physiologists lay out the law for them without enough clinical testing', as the Swedish doctor Curt Wallis argued.[7] For practicing physicians, desperate to find a cure for phthisis, pellagra, anaemia and other wasting afflictions, lamb blood transfusion seemed a promising way forward. Despite disappointments, many argued for continued clinical trials. It was necessary to keep on trying and trying again.

The quarrels concerning lamb blood transfusion are instructive. They illustrate the difficulties at the time in reaching a consensus about what should count as reliable medical proof. Traditional forms of medical knowledge production competed with new, science-based ones. Laboratory scientists and practicing doctors understood the sick body differently: why it was ill and how it could be cured. They worked in different social settings, with different means of gaining knowledge and assessing it.[8] This meant that neither group would readily accept the other's results as conclusive evidence. In addition, neither mode of medical knowledge production was, at the time, sufficiently developed for its arguments to be immune to criticism, and blood itself was poorly understood. We are on 'the wide field of conjectures, beliefs and hopes', as one German surgeon phrased it in 1874.[9] Lamb blood transfusion was therefore, in a sense, a *mystery*; it was not self-evident how its effects should be understood.

This chapter will not resolve the enigma, but it will illuminate the controversy. We will listen to the arguments for and against lamb blood transfusion. For this purpose, we will first visit three social milieus where medical knowledge was produced: the village doctors' surgeries, the urban hospitals and the physiological laboratories. These were settings where, in the 19th century, only men were considered experts; women were patients and sometimes nurses, and with very little say in what went on. In all three surroundings, thus, male professionals produced knowledge about transfusion, but in different ways, with different goals and means. No wonder that they – the practicing doctors and the experimental scientists – sometimes were not ready to accept each other's arguments and results.

We start by renewing our acquaintance with Oscar Hasse, a fine representative of what may be called 'bedside medicine'. How did he (and doctors like him) gain knowledge about disease and cure? How did their everyday practices colour their understanding of lamb blood transfusion?

Bedside medicine

Hasse was a private practitioner working in and around the town of Nordhausen in northern Germany. He took care of all kinds of medical problems and he sometimes performed transfusions, either in his clinic or in the home of his patients. He was then often assisted by a neighbour, 'an elderly gentleman of Nordhausen, not a medical man, but someone who had the advantage of having already frequently assisted Dr Hasse – a necessity for carrying out the operation with precision', reported a patient who, suffering from consumption, had asked to be transfused by Hasse.[10]

A local doctor, like Hasse, 'had to know the individual dispositions of his patients, their ways of life, and their joys and sorrows.'[11] His patients were mostly rural or small-town middle class. Most would have the means to pay for his services, and, if necessary, for the use of a lamb. Hasse's decision about whether to transfuse or not had to be negotiated at the sickbed with the patient and the family; a sometimes delicate situation. The patients' own descriptions of their condition and their wishes for treatment played an important role. 'These are unfortunately the downsides of the practice in general, that the doctor not only has to deal with the disease but has to struggle with various elements surrounding the patient!', a contemporary German doctor complained.¹²

Figure 22. A doctor at the bedside. Painting by Luke Fildes, 1891 (https://commons.wikimedia.org/wiki/File:The_Doctor_Luke_Fildes_crop.jpg).

A fair number of those inspired by Hasse's example were doctors in private practice. Their transfusion reports hint at a certain pride that a country doctor could invent a therapy adopted by hospitals across the world. That this was the case did, on the other hand, greatly annoy the celebrated Danish physiologist Peter Ludvig Panum. He had studied in Paris and Würzburg and worked with Rudolf Virchow and Claude Bernard. He was professor in Kiel but moved to Copenhagen at the start of the Danish-Prussian War. Panum had made many of the animal experiments that, from the early 1860s onwards, were used as evidence for the value of indirect transfusion with defibrinated human blood and, also, as proof of the dangers of species-alien blood.¹³ He now, in 1875 and 1876, published two quite sarcastic articles where he dismissed Hasse's (and Gesellius') evidence for the positive effects of lamb blood – it was totally fraudulent and misleading. Hasse, being a simple 'provincial doctor', had not, Panum claimed, understood the finer points of physiology but had based his suggestions on the sole but erroneous criterion of success at the bedside. The result was a hazardous gamble for the unfortunate patients.¹⁴

Panum was particularly upset about the many innocent 'village doctors' misled by Hasse and Gesellius. Naturally, they could not keep up with the complex developments in physiology and had therefore 'to an unfortunate degree been groping in the dark as to the indications for a transfusion'.[15] Their ignorance and 'misdirected ambitions' had fooled them into following Hasse's example, and apply the method for conditions – phthisis, cholera, leprosy, scurvy, melancholy, erotomania – that could never be helped by a transfusion. In this way, a dangerous 'epidemic' of lamb blood transfusion had spread across Europe, from Petersburg to Bonn, from Copenhagen to Italy:

> Hitherto unknown doctors have with the help of the reintroduction of the DENISian lamb blood transfusion achieved large local fame by establishing themselves as lamb blood transfusionists in villages blessed with lamb, where tens of phthisikers and other luckless patients have been transfused with the symbolic blood of lamb.[16]

Panum was right in that local doctors often did grope in the dark. Their patients' condition was frequently difficult to diagnose. It is worth recalling that average life expectancy in 1871, in for example Germany, was only thirty-seven years. Many children did not survive their first years but adults, too, had a hard time.[17] Local doctors performing a blood transfusion would first, they reported, have tried their usual therapies: enemas, hot and cold water cures, injections of ergotin, doses of opium and morphine, diets with meat or herb extracts – but to no avail. As a last resort, they tried blood transfusion. To them, the blood of a lamb seemed just as beneficial as that from a human being and less painful for the donor. And a lamb was perhaps (as Panum implied) easy for a village doctor to procure.

To gain knowledge about their patients' condition before and after a transfusion, these doctors employed quite simple means. They used their intuition and their five senses. They reported having checked their patients' temperature and pulse, sleep and appetite, as well as their urine and stools. They listened to the patients' breathing and heartbeats, looked at the colour of face, feet and hands, checked for urticaria, and smelled the patients' often quite unpleasant breath and sputum. In some cases, they used a microscope to assess the presence of red blood cells and albumin in urine, but they did not count the number of blood cells. In only one case reported by a village doctor was an autopsy performed; this was something that otherwise only took place in hospitals.

Hospital medicine

Local doctors sometimes also worked in nearby hospitals, and some performed transfusions there, too. In fact, most transfusions in the mid-19th century, including those with lamb blood, took place in hospital settings – in city hospitals, asylums, spas, and military hospitals.

The physicians and psychiatrists performing transfusions in these settings were certainly no unknown 'village doctors'. They were highly educated. Several were or would become professors, chief military surgeons, heads of clinics or mental hospitals. Panum was perhaps aware of this situation when he added that it, in no way, had been only ignorant provincials 'who with some enthusiasm had resorted to using animal blood [...] but also several renowned and undoubtedly honourable men'.[18]

These 'honourable men' were often explicitly supported by their hospital administrations and colleagues. Their transfusion attempts, be they with human or animal blood, were seen as important experiments. Results were reported in books, articles and dissertations. Medical societies across Europe and the USA held meetings and organized committees to debate transfusions. Thus, the mid-19th century transfusion experiments reflected professional ambitions within several medical communities (most noticeably perhaps within Italian psychiatry) and were not primarily individual whims.

The transfusion situation itself was an important occasion to communicate findings and observations, influence students and colleagues and even, as we have seen, impress royalty and the general public. This ambition to publicly inform others of clinical results was part of what we may call a 'hospital mode of knowledge production'. In contrast to bedside practices where the doctor's knowledge of their patients' condition was a kind of local and private property, the new ideal was communication. To further medical progress, and their own careers, physicians had to make their experience known to a wider medical community. Journals and meetings constituted a sphere where medical knowledge was presented and judged. Acclaim of peers within this larger, public domain was an endorsement of the doctor's position as an expert.[19] Meetings and conferences, and the medical press, were venues also for debates, quarrel and controversy; this certainly turned out to be the case for lamb blood transfusion.

How else did knowledge production in the hospital differ from that at the bedside? In the mid-19th century, physical examination (inspection, palpation, percussion and auscultation) had become routine diagnostic practice within hospitals (as seen, for example in Figure 23). Bodily functions, such as temperature, respiration and pulse were systematically measured and charted. These diagnostic practices were largely similar to what was used in bedside medicine. What differed were two things: the generalising ambition of the hospital mode of knowledge production and the nature of doctor-patient interaction.

With a start in the large Paris hospitals of the early 19th century, knowledge about diseases and appropriate means of redress was produced through careful clinical observation, classification of symptoms and diseases, and systematic recording of hospital statistics.[20] Such investigations were possible in hospitals with their many patients who could be observed for a stretch of time and, if they died, be subject to autopsy. Physicians could then correlate the signs and symptoms they found in the living patients with the structural changes they observed in post-mortem examinations. They could use surgical techniques to dissect the bodies and find exactly where the disease had been located. In this way, pathology became the foundation for a unified art of healing.[21]

In the 1850s, this localized theory of disease was radically revised by the work of the German physiologist, Rudolf Virchow. To him, the seat of disease was no longer the organ or the tissue as such but the cell; surgeons could therefore cut out the diseased cells without compromising the function of the rest of the body.[22] Sophisticated surgical interventions to treat disease by removing organs or parts thereof could now become standard elements of hospital medicine. They did not belong to the bedside doctors' repertoire since they required operating rooms, instruments and medically-trained assistants.

The nature of doctor-patient interaction changed, too, with the advance of hospital medicine. A culture of medical paternalism where the physician's authority reigned supreme characterized many 19th century hospitals. The 'previously shared knowledge about disease between patients and their physicians, so useful in forging a trusting relationship and negotiating therapeutic strategies [...] was shattered. For treatment, patients now became much more dependent on their physician's knowledge and judgment', a later historian summarized the situation.[23] Hospital patients were mostly poor or working class and many were illiterate, something that left doctors with great margins for what to do, how and when. Patient status was communicated in a technical language that most patients found hard to understand. At meetings and in articles they were made into 'cases' or became items in aggregate statistics of diagnoses and therapeutic results. This lack of personal doctor–patient involvement was sometimes regretted by physicians: 'Medicine [looks for] ... facts, it has become objective. It does not matter who is at the bedside, the sick person has become a thing', a German doctor protested in 1870.[24]

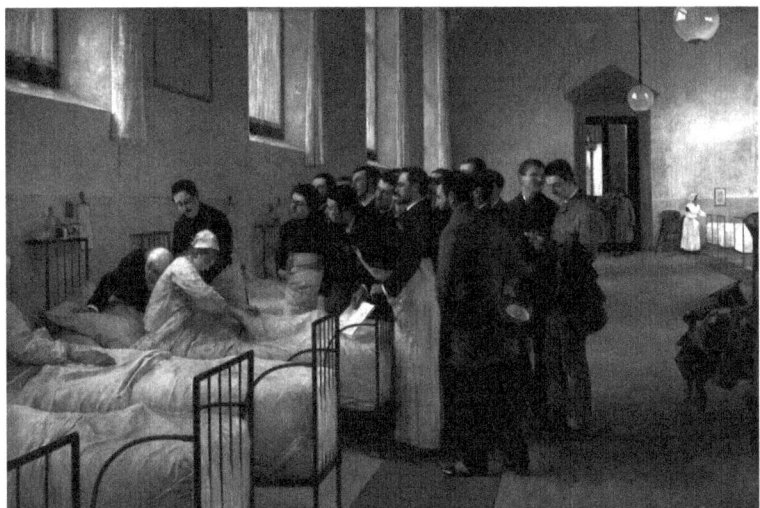

Figure 23. A visit to the hospital. Painting by Luis Jimenez Aranda 1889 (https://commons.-wikimedia.org/-wiki/File:La_-visita_al_-hospital_de_-Luis_Jim%C3-%A9nez_-Aranda.jpg).

Hospital doctors of the 1870s could perform quite advanced surgery with the help of anaesthetics and Listerian antiseptics. Such interventions were now less painful and more likely to succeed. Otherwise, and judging by their transfusion accounts, they used much the same remedies as the local practitioners. Homeopathic and hydropathic treatments were common. Doctors made turpentine enemas, used ether injections, applied mustard plaster on breast or legs, administered lead lotion, chinine, eucalyptus tincture or Carlsbad waters. Extracts of meat, malt and milk were given and there was, 'a vogue for the use of alcoholic beverages as stimulants'.[25] Transfusion patients were served red wine, champagne, port or milk mixed with brandy, sometimes before but most often after the ordeals of the intervention.

Hospital doctors were university trained. They had a fair amount of theoretical knowledge. Some – especially the Italian alienists – held scientific ambitions; several had extensive research activities. There was a strong sense that clinical interventions should be based on science. But blood was in many ways a mysterious fluid. Theoretical knowledge of its components and their function in the body was still limited. This did not prevent many of those doing a lamb blood transfusion from backing up their results with various, more or less well-grounded physiological arguments. But *real* science, their opponents argued, was not done by clinicians in the hospitals – it was performed somewhere else.

Laboratory medicine

We therefore move to a third place for the production of medical knowledge – the laboratory. For example, to the one of L. Lesser, a physiologist in Berlin. Here he is, in 1874, giving a lecture to members of the Obstetrical Society of Berlin:

> Permit me [...] for a few moments to take you away from the bedside into the experiment-room of a physiological laboratory. The experimental physiology of the blood will, I trust, give you a better answer to many obscure questions in the study of the replacement and saving of blood, and you may also find in it a more certain footing for your medical treatment than in all the casuistry hitherto so prevalent in the science and art of therapeutical transfusion.[26]

Beginning in the mid-19th century, well-endowed physiological and pathological laboratories for research and education were established across the German-speaking world. They were to be found in, for instance, Heidelberg, Greifswald and Zürich and, on a more 'grandiose scale', in Vienna, Berlin and Leipzig, as an impressed French medical emissary reported in 1870. Nothing like it existed in France or even Great Britain.[27]

Work in these laboratories would, as Lesser and others claimed, put medical treatment on 'a more certain footing' than mere clinical experiments would allow. The local village doctor could in an emergency hardly deliberate on the solubility of blood-corpuscles or whether to use a direct or indirect method of transfusion, argued a writer in the *Medical times and gazette* in 1874. Instead, 'these and the other points involved should be decided for him by the clinician, whose labors, it seems to us, should be based on the results of the physiologist.'[28]

Figure 24. The histological laboratory, San Lazzaro Asylum, Reggio Emilia. (Courtesy of San Lazzaro Asylum Archive, Reggio Emilia. Album A7 photo n. 11, C d 4.12 immagine 013.)

Physiological experiments, thus, were thought to give the solid knowledge about tissues and cells needed for hospital medicine and, eventually, bedside care. Such information was, in the case of transfusion, largely based on animal experiments; it was assumed that their results were valid also for how the human body would react.[29] Laboratory manuals and accounts of the time provide detailed, sometimes gruesome, insights into how the scientists worked, their techniques and their treatment of the animals.[30] Landois, for example, whom we have met earlier in this chapter, subjected large numbers of animals to often painful experiments and careful observations. He employed a modified Aveling transfusor to move blood from one animal to another, a kymograph to measure blood pressure in the transfused animal, and various contraptions to fixate and inject the frogs on which numerous experiments were made.

By the early 1870s, laboratory medicine had, it was argued, reached the conclusion that only blood from the same species could safely be used in transfusion. But then, suddenly in 1874, practicing doctors across the world claimed success against various diseases using lamb blood transfusion. To the astonished physiologists, it seemed as if 'everything that [they] had shown was built on loose sand and destined to collapse in the face of a rapidly gained practical experience', the Swedish physician Warfvinge remarked.[31]

To this challenge, the experimental scientists reacted in two ways. First, with verbal counterattacks. For Panum, the struggle was now between 'crude, unscientific, uncritical empiricism', on the one hand, and 'scientific medicine that makes use of physiological, pathological and pharmacodynamic experiences and facts', on the other.[32] He was echoed by the Swedish pathologist Rossander:

> For the sober and sceptical observer, some miraculous cures are not proof enough; he wants to see clear reasons and arguments, he demands for the solution of such great questions, not simply some more or less successful 'cases' but a scientific foundation for these.[33]

In text after text, the 'calm, conscientious' and 'sceptical' scientist was set against the uncritical and hectic, even maniac, advocate of lamb blood transfusion.[34] Hasse was the prime target; he felt the attacks quite keenly and personally. He accused the physiologists of vilifying him to scare doctors away from performing potentially life-saving transfusions:

> Our most important physiologists with all the force of their authority, with the sharp weapons of their minds, with all the equipment of their physiological laboratories, and with their numerous auxiliary troops of assistants and pupils, use this erroneous image [of lamb blood transfusion] to make the simple provincial doctor worried and afraid.[35]

Secondly, the physiologists set to work to produce more laboratory evidence for their case. Professor Landois soon demonstrated, with a new series of experiments, the perilous effects of lamb blood transfusion, while Professor Ponfick in Rostock found that red blood cells of species-alien blood dissolved in the receiving organism's blood plasma. Its haemoglobin would then excrete into the urine to cause haemoglobinuria, a potentially fatal condition, and the kidneys would get overworked.[36] Once again, it seemed that science, as one observer phrased it, had dealt 'a crippling blow to the troublesome direct lamb blood transfusions'.[37]

Still, the verdict was far from clear. For a clinical intervention, such as lamb blood transfusion, to be considered beneficial and safe, it seems that at least three conditions have to be met. Firstly, it should, if possible, be based on theory and scientific evidence. This was a new idea in the 19th century and, as we shall see, not without its problems. Secondly, it should make the patient better, also in the longer perspective. And thirdly, it should be safe and not cause undue harm. If and how these conditions were met was, at the time, a matter of contention.

I find the arguments presented for and against lamb blood transfusion worth discussing in some detail. They signal a genuine uncertainty, not only about the effects of this particular intervention but, more generally, about how different kinds of medical evidence should be assessed and compared. Hospital and bedside based doctors tended to favour clinical experience and distrust animal experiments; physiologists thought quite the opposite. Still, the evidence was far from clear-cut; there were doubts on both sides as to the relevance of their respective arguments. Or, as noted by a somewhat disillusioned observer: '[T]ransfusion has [recently] become a favourite object of physiologists, experimental pathologists and many surgeons. The [...] literature has risen to an enormous height, but with it also the confusion'.[38]

Laboratory experiments contested

To sort out this confusion somewhat, I will first summarize the critique against the merit of animal experiments. I will then consider the other side: the arguments for and against the merit of clinical experience. A somewhat inconclusive situation will emerge. But perhaps the statistical treatment of available data may help in reaching a consensus? An unfounded hope, as we will see.

I start with the laboratory experiments that, physiologists argued, dismissed lamb blood transfusion as useless and dangerous. But some lamb blood proponents did not accept these results as evidence. They questioned how the experiments were carried out and their relevance for clinical practice. For example, the Austrian military surgeon, Neudörfer – whom we have met as a supporter of lamb blood transfusion in war and peace – argued that the serum used by Landois was an artificial product that dissolved red blood

Figure 25. A meeting of the Swedish Medical Society in 1879. Illustration by Carl Larsson (Ny Illustrerad Tidning, December 27, 1879, 401).

cells much faster than what would happen in the human body. Thus, conclusions drawn from his experiments might not be relevant for clinical practice.[39] Roussel, too, was critical of the particular transfusion instrument (Aveling's, not his) used in Landois' experiments:

> This physiologist transfused dog's blood to cats, frog's blood to rabbits; he operated, and he showed his results with the patience and detail characteristic of the Germans. [But] this long study is tainted with inevitable errors produced by an unreliable transfusion method. He can affirm, neither that the blood used has not been altered on contact with the air, nor that the blood itself has retained all its qualities and physiological force.[40]

Other commentators were sceptical about inferring conclusions from animals to humans, and from healthy individuals to sick ones. Even the physiologist Emil Ponfick, who had demonstrated that dissolving lamb blood cells caused a potentially fatal haemoglobinuria in the recipient, was somewhat reticent. He warned against drawing strong conclusions from animal studies. The British journal, *The Doctor*, summarized his reluctance:

> In terminating his remarkable work [...] Ponfick declares that he in no way intends to infer from what he has noticed in some animals, especially in dogs, what would take place in man; he also does not desire to resolve the most important questions in practical medicine by considerations solely based on observations made on persons in good health. His aim in writing these lines was simply to augment our knowledge as to the influence of transfusion on the animal economy.[41]

Another contention concerned the *amount* of blood transfused. Laboratory animals, in contrast to human patients, often received quite a lot of blood. This issue was raised by a number of critics. One was the French physician, Jean-Cyprien Oré, in Bordeaux, a pioneer of anesthesiology with a long-standing interest in transfusion. To those who argued that animal blood transfusion was both useless and dangerous, he countered that this depended not on the *kind* of blood transfused but on *how much*. He himself had transfused dogs with blood from various animals with no danger to the animal and with no destruction of the red blood cells transfused. The adverse effects encountered by others were, he argued, due to them administering an overabundance of blood given the weight of the recipient. Thus, it was not surprising that these physiologists would encounter bloody froth and urine, followed by the death of the transfused animal.[42]

Other lamb blood defenders agreed. A blood transfusion to a human patient would only introduce some ten to twenty per cent of what the scientists gave their laboratory animals. This small amount of transfused blood would then act as a drug, not as a poison. The Swedish pathologist Rossander (though a sceptic to lamb blood transfusion) somewhat cheekily remarked:

> The physiologists may experiment with their poison, inject dogs [...] with large doses thereof, but in small doses any poison may under certain circumstances become a medication. If you inject enough morphine or strychnine into an animal, you will kill it; this does not prevent both from being excellent remedies for humans. The same may be the case with haemoglobin.[43]

There were other problems, too, with inference from animal experiments. 'The experiment only teaches us how animals fare', the German doctor, Jahn, argued. Despite being favourable to experiments, he noted that 'no experiment gives us information about the success of transfusion in various internal human diseases that we cannot produce experimentally in animals'.[44] This was a valid remark, most relevant perhaps, for the cases of insanity treated by the Italian alienists. I have found no account of transfusion experiments on, for example, mad dogs, from which conclusions to human mental patients could have been drawn.

Thus, there were question marks concerning the relevance of laboratory evidence for clinical practice. But, on the other hand, how reliable was the clinical experience? How beneficial was lamb blood transfusion, in the short and the long run? Here, too, the data was partial and confusing, leading to contrasting views on its merit as evidence.

Clinical experience contested

Two main types of methods were at the time used to prove, or disprove, the efficacy of a clinical intervention. The first was close observation and comparison of data from individual cases, the second statistical analysis of the information from a large number of cases. Both methods were referred to in the debate and both were beset with problems.

Doctors who had tried lamb blood transfusion, seemed quite eager to report on their experiences, both positive and negative ones, and sometimes in quite long-winded detail. Many claimed amelioration or full recovery of their patients. The Swedish physician Lamm summarized the situation, as he saw it in 1875, as follows:

> Genus homo can, according to what experience has shown us, quite well support immediately transfused blood from the sheep species [...] Also after necessary discount of the authors' accounts, it seems that one cannot doubt the good effects on humans of lamb blood, that is, of heterogeneous blood in toto. I have noted no deaths by poisoning from the transfusion of such blood.[45]

Such results, the German doctor, von Cube, maintained, were 'an example of the favourable effect of this operation, although it may at times be incompatible with the results of scientific research'.[46] As Jullien hinted above, doctors being close to their patients saw the worth of the intervention differently than did scientists who were 'disdainful of the sick and confined in the heights of physiology'.

The clinicians' case reports were, however, not always easy to interpret. The Swedish doctor, Ivar Svensson in Oskarshamn, who had tried transfusion with both human and animal blood but with little success, saw most case reports as expressing wishful thinking among doctors and patients hoping for a miracle cure.[47] Even supporters, like von Cube, had to admit that many positive accounts were based on such unfounded assumptions, even speculations, that they could not really promote the cause.[48] The reader may remember the puzzlement of the Dresden physicians, Fiedler and Birch-Hirschfeld, who made careful comparison of their own (unsuccessful) and Hasse's (successful) transfusions to phthisis patients. No relevant parameter seemed to account for the difference in outcome. Although they themselves were against what they considered to be a painful operation, their conclusion was to wait and see what future experience would bring.[49]

One particularly contentious aspect concerned how much lamb blood was actually transfused, an issue that I referred to in chapter 4. For Panum, it was most likely that only very small amounts of blood had been transfused in each case; that was, to him, probably the only reason why no patient had died from a lamb blood transfusion.[50] For some lamb blood supporters, on the other hand, like Oré and the Italian psychiatrists, it was precisely this manoeuvre – the transfer of only small, but repeated, amounts of lamb blood – that allegedly made for its success.

Another point of contention was that most published case reports were quite poor in information. Lamb blood transfusions may have been called 'experiments' but they were not, properly speaking, clinical trials, critics argued. There was simply not enough data presented. 'Innumerable experiments have been performed but without any precise settling of the question and without any strictly scientific method', Lesser argued.[51] Some Italian scientists were extremely critical of the cases presented by, for example, Ponza and Manzini and Rodolfi: their accounts were allegedly short of useful clinical data, with no blood counts and no systematic descriptions of the patients' weight, temperature, the state of their kidneys and other clinical data.[52] Landois had a similar critique of Hasse's reports:

> But one aspect in particular has always remained incomprehensible to me: why has this modern Denis not one single time used a prick of a needle to get a small drop of blood from his patients for the microscopic examination of for how long time the lamb cells are still visible in [the human recipient's blood]? That would surely have made him change tracks.[53]

Thus, no real conclusions, critics argued, could be drawn from these so-called experiments. It was impossible to ascertain why some people felt better after a transfusion while others did not improve or even died. Perhaps it was not the transfusion itself that had helped the lucky ones but something else? Phthisis patients, for instance, often recovered spontaneously, as did some early stage pellagra patients. Also, many of those transfused were poor people who had left their insalubrious surroundings; they probably gained strength more from the food, rest and care in the hospital than from the transfused blood.

A further important question concerned how one should define 'success'. Some transfusionists counted an only temporary improvement as a positive result. The increased appetite, the good night's sleep and the improved digestion could, they argued, help the patient recover and be ready for other treatments; thus a transfusion was worth trying.[54] For other physicians – like Dr Mayer, a private practitioner in Munich – the intervention was a humanitarian, and thus beneficial, act in an otherwise hopeless situation:

> It's more comfortable, of course, to let [the patient] die quietly so as not to torture him any more, as the popular expression goes, but it is inhumane and as a doctor I hold to the obligation to prolong, even if only by 5 minutes, the life of a person who has confided in me for help.[55]

The problem with statistics

When the number of case reports began to pile up, there was need for an overview. Quite a number of statistical evaluations of *human-to-human* transfusions had already been made. Martin had published one in 1859, Oré one in 1868, von Belina one in 1869, Marmonnier one in 1869, Sacklén one in 1870, and Gesellius one in 1873.[56] Now, it seemed useful to compile and evaluate statistics also about *lamb* blood transfusion.

In principle, such compilations could be instructive. Still, there was a major problem. Given that the case reports on which they were based were so incomplete, they were not easy to systematize and compare. Here is Dr Jahn again. He was, as noted above, sceptical of existing laboratory studies but he also questioned the possibility of drawing conclusions from compilations of extant cases. He had found a number of difficulties in the statistics available in 1874 (when the figures only concerned transfusions with human blood):

These statistics list a large number of experiments made on sick people that seem to possess a higher value and provide better proof than those done on animals; but with this advantage come some very significant disadvantages. None of the observed cases is based on such simple and precise questions as we demand of an experiment, and so many other circumstances are involved that the separate cases are of no use. We may seek to counter this inconvenience by compiling a large number of cases and comparing them with one another in order to eliminate the incidental coincidences attached to each individual case. No matter for what purpose the statistics are to be used, to answer our questions with certainty requires large series of cases, much larger than what the previous literature on transfusion has been able to provide.[57]

Hasse clearly understood the importance of getting a large set of detailed data. Already in April 1874, at the Congress of the German Surgical Society, he distributed a questionnaire, asking his colleagues to report details of their upcoming lamb blood transfusions with indications, procedure and results, and send the information to him. He obviously expected a high number of replies.[58] Of this initiative, however, no more was heard (except that Panum made fun of it).

Others, however, compiled statistics from published lamb blood transfusion reports. In 1876, the French doctor, Jean-Cyprien Oré, published an update on his 1868 human-to-human transfusion statistics. He argued that his compilation of animal blood transfusions as well as his own animal experiments (see above) had shown that lamb blood transfusion was both useful and safe. He based his argument on 154 reported observations of lamb and a couple of cases of calf blood transfusions to humans. Especially instructive were, in his view, the Italian cases where very little blood had been transfused. The Italian alienists had reached, Oré thought, the most remarkable results with only few strong side-effects. He concluded his overview by affirming that: 'once more, the clinic has confirmed in a striking manner the results established by experimental physiology'.[59]

Two other attempts to compile and analyse case reports are worth noting. They were made by the German, Landois, and the Swede, Warfvinge, in 1875 and 1876, respectively, with quite different goals in mind. Landois wanted to show the danger and uselessness of lamb blood transfusion while Warfvinge wanted to stress its possibilities. But none of them was capable of doing a sophisticated statistical analysis and their data were, as indicated above, uncertain and incomplete (the same can be said of Oré's report). Warfvinge's results were in favour of lamb blood transfusion but quite weakly so and only in cases of anaemia. Meanwhile, Landois did not discriminate between direct and indirect transfusion when counting the varying results of the intervention (death, favourable, unfavourable, tentative), and did not publish any percentages. Later commentators found that his figures actually went against his conclusion that lamb blood transfusion was dangerous. Only twenty-nine per cent of the lamb blood patients died compared to fifty per cent of those transfused with human blood.[60]

Of particular interest are the compilations made about lamb blood transfusions to phthisis patients, the subject of chapter 4 above. In 1876, the Swedish doctor, Curt Wallis, counted sixty-five international cases of such transfusion. Of these, nine patients had died, thirteen had improved, and the rest (forty-three) had experienced no improvement or their fate was unknown.[61] To this list, I (with the benefit of being able to scan the international literature with digital methods) can add thirty-five cases not included in Wallis' account. Of these, six had died, twenty improved, and nine got worse or their situation was uncertain. Overall, thus, a third of the phthisis patients, some of whom had been in a very sorry state before the transfusion, were reported as improved. At the same time, two thirds were most likely not and the positive estimates are highly doubtful, given the scarcity of medical information and the very short time, in some cases, between the transfusion and its reporting.

Not surprisingly, Panum was sceptical about such compilations of clinical cases, be they of human or lamb blood transfusion. He considered them incomplete and the cases so heterogeneous in their indications that they were useless for all practical purposes.[62] The Italian physician and psychiatrist, Cesare Lombroso, agreed. Quite different diseases had been grouped together; benign illnesses had been labelled as incurable ones to show the wonders of a transfusion; deaths occurring after some time were not reported; the impact of deficient instruments was not taken into account, nor were the possible differences registered between transfusing robust young patients and more frail, older ones, etc.[63] Roussel, who was sceptical to everything except his own instrument, refused to compile any statistics at all, not even of his own numerous transfusions:

> Whatever others may say, statistics have absolutely nothing to do with medicine, because it is easier to find two identical leaves than two similar

human beings in terms of their constitution, their predispositions, their current malady, their susceptibilities and their reactions when being exposed to the same medicine.

All my transfusions are different in terms of their causes, their doses, their reasons, their effects: I accept no arithmetic whatsoever and I do not answer to any demand for percentages.[64]

***

Thus, we are still left in the dark about the medical evidence concerning the pros and cons of lamb blood transfusion. Did it work? There seemed to be no definite verdict, neither from the laboratory nor from the field.

But we should not forget the third condition for accepting or rejecting a therapy: its harm. Given the pain and uncertainty involved in a lamb blood transfusion, one may also legitimately ask: *Was it worth it?*

This question will be discussed in the next chapter.

7. Transgressions

> Imagine lying on a board, your legs tied together, a cloth covering your eyes. Your neck hurts, something is protruding from it. Then you feel fainter. You sense something moving jerkily next to you, you hear screams. A few minutes later, there is another sharp pain by your throat. You are moved to somewhere else, released. You stumble awkwardly onto the grass. You do not know it, but you have just lost 150 ml of your blood.

In 1879, Professor Peter L. Panum was attacked. The aggressor was the Danish Society for the Protection of Animals [Foreningen til Dyrenes Beskyttelse]. It had just published a translation of a pamphlet, in English called *The Torture Chamber of Science*, by the German anti-vivisectionist, Ernst von Weber, containing lurid accounts of suffering animals in scientific experiments.[1] To the activists, Panum was a major Danish representative of such cruel practices. Now they vilified him in pamphlets and newspaper articles, and he replied with passion. Animals were less liable than humans to feel pain, he argued, and animal experiments were needed for the progress of science and medicine. To prove his case, he made the following interesting comparison:

> A couple of years ago, when lamb blood transfusion was en vogue, no one doubted that a doctor was entitled to sacrifice a lamb if he had an ever so weak hope of thereby saving or prolonging (if only perhaps for a short time) a human life [...]
> Should not a physician, who has, and must necessarily have, the right to treat his sick fellow beings according to his own judgment and conscience

and without interference, also be allowed to, without interference, decide over an animal's life and health in the interest of humankind?[2]

It is not likely that Panum had changed his mind about lamb blood transfusion since we last met him. Still, he used it as an example of a presumably acceptable procedure to defend his laboratory practices. This raises questions: How were the ethics of animal blood transfusion perceived by contemporaries? Was it seen as something banal – or as an improper transgression of natural and cultural boundaries, a cruel use of animals and a dangerous experiment on vulnerable patients? And how did this compare to the use of animals in laboratory experiments?

As Panum indicates, there was at the time little institutional control of therapeutic and experimental practices. An authoritarian culture reigned in hospitals, asylums and research laboratories. No legally binding ethical guidelines helped doctors decide in morally tricky situations at the sickbed, no explicit rules of conduct guided physiologists in their experiments. Everyday practices were a matter of individual conscience and situated judgement.

So, to answer the question, 'was it worth it?', we have to examine how physicians and scientists in the mid-19th century reasoned about the morality of their transfusion practices. Was the progress of medicine worth subjecting patients to a perhaps dangerous intervention, and animals to painful experiments? This was a complex issue with no clear consensus.

Using animals

There is an interesting paradox, apparent in today's discussions of organ or cell transplants from animals to humans (xeno-transplantation) but relevant also for the 19th century: the animal from which tissues or organs are to be taken should be sufficiently *similar* to humans for it to be medically possible for our bodies to accept the transplant. At the same time, the animal should be sufficiently *dissimilar* to us for us to consider it ethically acceptable to exploit it, kill or molest it, and make it suffer for our sake.[3]

Deliberations in the 1870s about the use of lamb blood for transfusion almost exclusively focused on the first, *physiological*, question: how similar or different is human blood to that of other species? What happens in the human body when non-human blood is introduced and how beneficial or dangerous is it?

Opinions differed. They ranged from finding species-alien blood quite similar to humans' and thus useful (as long as its blood cells were smaller than or the same size as those of human blood), to seeing it as so different that it was poisonous for the receiving organism. Then there were intermediary ideas. The physiologist, Landois, found that blood from animals of the same taxonomic family, such as fox and dog between which he performed reciprocal transfusions, was nominally similar enough.[4] Hasse presented a developmental version of this evolutionary idea in his reply to Panum in 1875. It may be so, he speculated, that the blood of the little lamb is healthier for us than that of the full-grown sheep since young animals are closer to humans than are older ones. This view was ridiculed by Panum as yet another absurdity peddled by an ill-informed village doctor.[5] Still, Hasse was not alone. He was most likely influenced by (but maybe misrepresented) the German scientist Ernst Haeckel's 1866 biogenetic law, regarded as valid until the early 20th century. Commonly stated as 'ontogeny recapitulates phylogeny', it theorized that the stages, which an animal embryo undergoes during development are a chronological replay of that species' past evolutionary forms.[6] Hence, a young lamb could be hypothesised by Hasse as being less specifically a sheep than an older one, and its blood therefore closer to that of humans.

The parallel question about the *moral* acceptability of making animals suffer for our sake was not explicitly discussed by 19th century transfusionists. Perhaps this was only natural in a largely rural society where animals were kept for their usefulness for humans and not as pets, and where they were slaughtered often in full public view. Still, many doctors doing animal blood transfusion tried to minimize the pain and discomfort of their lambs. Manzini and Rodolfi recommended transfusions from the animal's vein rather than its artery since that was less painful, and Neudörfer chloroformed his sheep 'for humanitarian reasons'. He wanted to avoid frightening the animal, and also prevent it from scaring the patient with its human-sounding cries and sobs.[7] Several physicians underlined that the lamb fared well after the operation: it ate with good appetite and jumped happily about in the field. Hasse gave it nutritious food: grains and soaked peas, and once even took it into his apartment to recover.[8] Still, some animals bled to death (and sometimes they were destined to end up in a stew anyway).

A different kind of treatment awaited animals in the laboratory. Ever since Harvey in the 17th century, physiologists had made painful animal experiments when trying to understand the circulation of blood or investigating the possibility of transfusion. Magendie, Bernard and Brown-Séquard were 19th-

century pioneers in France; German scientists followed suit. I here focus on two of them, since they were involved in the lamb blood controversy: Panum and Landois. Panum, in the early 1860s, emptied his experimental dogs of almost all their blood before introducing the blood of others, sometimes from dogs, sometimes from other species. He also injected them with solutions of rotten meat, leading to painful and lethal effects. Landois in the 1860s and 70s transfused dogs, rabbits, cats and various other animals with blood from lamb and other species, with poisonous blood, or with blood whose blood cells had been killed off by heat. He nailed frogs to boards for them to be transfused and cut up. He subjected dogs to blood from carbon monoxide intoxicated rabbits, and he, Panum and others starved their experimental animals for days before depleting and transfusing them. Animals died in these experiments or were killed for the scientist to investigate the status of their organs and blood.[9]

From the mid-19th century onwards, physiologists could use ether or chloroform to alleviate pain for the animals, but I have found no such usage reported from the transfusion experiments. Some animal experimenters did not employ anaesthesia since the very point was to study pain reactions. One example is the Italian physiologist, Paolo Mantegazza. He was a most vocal critic of lamb blood transfusions to humans but did not hesitate to subject his experimental animals to cruel tests, for example for his 1880 study, *Fisiologia del dolore*, on the physiology of pain.[10]

Thus, there was a clear difference in how animals were treated at the bedside and in the laboratory. I interpret the relative care that transfusing doctors took of the lamb as a sign of them seeing it as somewhat of a collaborator in the transfusion endeavour. The lamb was an instrument of transfusion, the source of the necessary vital fluid. It should be handled with care. In some cases, it was scheduled for re-use a second and a third time; often it should be handed back to its owner who wanted it in good shape. Sheep were precious, not only for their blood, but for their meat, wool, milk. They must not be wasted.

The situation was different in the laboratory, although it seems that lamb, perhaps for economic reasons, were not experimented on in transfusion studies (but their blood was given to other animals).[11] Laboratory practices of the 19th century built upon what medical historians, Cunningham and Williams, describe as a profound change in sensibility on the part of scientists. 'The live animals had to be transformed into and be perceived as simply a neutral object of scientific investigation and not as a perceptive pain-feeling fellow creature being submitted to torture'.[12] Panum's preferred experimental ani-

Figure 26. Experiment on a living dog, according to The Torture Chamber of Science (von Weber 1880,1).

mal, the dog, had no particular value, economically or emotionally. It could not be milked or eaten. Stray dogs were ubiquitous, as was another often-used laboratory animal, the rabbit. The animals differed in age and appearance, but this seemed unimportant for experimental purposes. Panum and Landois occasionally noted the colour of the dogs used, their size, if they were young or old, and sometimes, mysteriously, their breed: a Fleischerhund, a Windhund, a Jagdhund, a Pudelhund... Other animals used (cats, rabbits, frogs, guinea pigs) were even more anonymous. Still, we are far from the standardized, commercial lab rats of the 20th century.[13]

I see the animal experiments of the 19th century as expressing what anthropologist, Philippe Descola, calls a *naturalistic* ontology of non-humans, which is another way of seeing the similarity and difference that I noted in the beginning of this chapter.[14] Since the animal body in the laboratory stands for the body of the human patient, it must be similar enough to serve that purpose. This shared physicality between animals and humans guaranteed the transferability of results from animal experiments to humans even if this sometimes was contested, as we saw in a previous chapter.

On the other hand, if animals are to undergo painful, degrading or lethal procedures, they must be *different*. They must have a lower status than humans in terms of ethical dignity since they are supposed to lack a humanlike interiority, what we call a mind, soul or consciousness.[15] Hence, we may, without raising moral concerns, use them, even in cruel ways, as a substitute for humans to produce general physiological knowledge.[16] Their bodies can be carved up, their arteries opened; they can be starved, poisoned and subjected to depletions and injections.

The use of lamb for blood transfusion to humans was, it seems, also based on a naturalistic ontology. The animal's blood was supposedly similar enough to ours. At the same time, the lamb was seen as lacking human subjectivity, the ability to think symbolically and the capacity to dream. Still, it was often treated with care. There was even a view that the animal, just because it was different, might have *better* blood than humans since it lacked our problematic interiority. We can return to the very first transfusions from non-humans to humans in Paris in the 1660s. Their initiator Jean Denis was convinced that the blood of animals was physiologically superior to human blood because animals were morally less disordered. He elaborated this point in a published letter about his first experiment:

> It is easy to judge that the blood of animals must have less impurity than that of men for debauchery and derangement in drinking and eating are not as common as among us. The sorrows, the worries, the fits, the melancholies, the anxiety and generally all the passions that are so many causes of the troubled life of man corrupt the substance of his blood. Instead, the life of the animal is much better regulated and less exposed to these miseries, the dreadful consequences of the sins of our first father.[17]

Experience shows, Denis continued, that it was rare to find 'bad blood' in animals whereas human blood was inevitably corrupted – the result, he reiterated, of man's fallen state.

This religious argument for the moral superiority of non-human blood was absent from the transfusing physicians' 19th century accounts. But it had an upshot in their warnings against using human blood: it might be corrupted by alcoholism, syphilis, gout, or other dangerous afflictions. Lamb blood was different – and healthier.

Still, was it not too different, too alien? Was it not morally and ethically unacceptable to subject patients to the experience of getting such *strange* blood into their bodies?

Crossing boundaries

To a Mary Douglas-inspired anthropologist, an animal in the sick room is an example of 'matter out of place'. Sheep are outdoor things that belong to animal pens, not indoors in hospital beds. Their presence there means a blurring of established cultural boundaries; therefore their blood becomes an ambiguous fluid: dirty, dangerous and disgusting.[18]

Doctors in the mid-19th century, however, quickly dismissed the issue of disgust, if they brought it up at all. The oxygen-rich blood from the lamb's artery was considered to be *natural* blood and was therefore, 'despite its disgusting animality [...] much better than human blood from the veins', Gesellius argued, and many agreed.[19] For example, Robert Barnes, a leading British gynaecologist, who assisted Aveling in the first British lamb blood transfusion (described in chapter 4 above):

> To supply an answer to the vulgar dread that with the blood of animals some noxious vital principle may be imparted, it ought to be enough to remember that man lives upon the flesh and blood of animals; and that it cannot matter whether lamb's blood be taken first into the stomach or directly into the veins.[20]

And, in an interesting twist, Barnes added elsewhere: 'No one would maintain that the blood of animals might not be taken into the human stomach, whilst the idea of swallowing human blood excited horror and disgust.'[21]

Barnes' colleague, Henry M. Madge, the secretary of a committee to evaluate different forms of blood transfusion, nevertheless, did acknowledge the possibility of nausea. To some people, he said, there may be 'something repulsive in the idea of bringing an animal into the sick-chamber and of mixing animal with human blood'. It was thus not simply a question of 'taking lamb in another form'. To avoid shocking the patient, Madge suggested that the physician use the indirect method and obtain the lamb's blood in an adjoining room to prevent the patient from seeing its animal origin.[22] Also the Italian physician, de Cristoforis, anticipated fear, apprehension and protest, particularly among his female patients, when seeing the bleating and trembling animal at close sight; he therefore opposed its use.[23]

Judging from the published reports, however, there was little such squeamishness. No patient is known to have expressed revulsion towards getting animal blood into their veins. On the contrary, Hasse reported, they begged him to give them this new medication and sometimes more of it than he thought fit. Hasse's direct method, used in the great majority of lamb blood transfusions, meant bringing the lamb and the receiver very close to one another. 'The human hand should be around the neck of the lamb', one physician recommended.[24] Still, I have found no reports of the shock reaction anticipated by Madge. Not even in the (unique) case when a large dog was used instead of a lamb (it being Easter time and no lamb was to be found). That transfusion had to be discontinued because the animal – not the patient – was too unruly.[25]

One may speculate about why the patients did not react with nausea or disgust and refused the transfusion. One obvious reason is that they were too ill. Many were unconscious, close to death. Another possible explanation may be that they just did not understand what was suggested to them. This brings up the issue of what today is called, 'informed consent'.

Accepting transgression

On July 6, 1874, the Turin newspaper, *Gazetta del Popolo*, published a denunciation of four lamb blood transfusions recently performed in the mental hospital of nearby Alessandria. Its author was Professor G. S. Bonacossa, a prominent Italian alienists, now in his seventies. He was upset. He would never, he stated, have permitted such an experiment in his asylum in Turin. It was contrary to the principles of humanity and medical prudence, and that for three reasons: (1) it was useless for the purpose of healing madness; (2) it had not been proven to be without danger; and (3) it was not allowed to perform such dangerous operations without the consent of the sick themselves unless there was an imminent danger to the patients and an almost certainty of restoring their health.[26]

The next day, the newspaper carried a response to Bonacossa. It was written by his Turin colleague, Professor Pacchiotti. He was one of the four psychiatrists behind the transfusions in Alessandria. Pacchiotti argued, first, that nobody thought that a transfusion would cure the mentally ill; it could, however, better than any other remedy, improve their anaemic state. Secondly, he considered the operation to be without danger and added that, if you only did what established surgeons accepted, there would be no progress in medicine. As to the third objection, concerning the necessity of having the patients' consent, Pacchiotti was more evasive:

> Yes, when they can give it. But how many operations are not done on children to save them from death! How many sick people do not accept an operation without having an exact idea of it! How many operations are not made suddenly after serious accidents when it is a question of saving the life of a man! And then again, the four mentally ill [in Alessandria] let themselves be transfused with the blood of a lamb, being quiet like lambs.[27]

It should perhaps not surprise us that patients in 19th century asylums, clinics and military hospitals submitted to whatever their doctors suggested. An authoritarian culture reigned, sometimes with militaristic overtones – what medical historian Andreas-Holger Maehle calls 'medical paternalism'. The patients' position vis-à-vis their doctors was a weak one. 'The ever-widening knowledge gap between medical experts and patients, and the increase in available diagnostic and therapeutic methods of hospital medicine, gave doctors more and more authority in decision-making', Maehle notes.[28]

Some reports of lamb blood transfusion do mention that a transfusion was suggested and accepted. For example, the Swedish doctor, Ivar Svensson, notes that his female patient – being 'as forbearing and compliant as could be imagined' – agreed to whatever he suggested, including an experimental lamb blood transfusion.[29] In most other accounts, however, nothing is said about patient consent; it was an implicit or silent matter. Hospital doctors conferred with colleagues or superiors but not always with patients or relatives; these were generally less educated and from a lower social class. Private practitioners, catering to a more well-to-do clientele, seem to have been more aware of the need for consent, perhaps because the transfusion entailed added costs for the patient. There was a practical need for communication and negotiation.[30]

How consent was obtained differed. In the mental hospital of Brescia, patients were induced to cooperate with the help of various delicacies. In Cincinnati, Dr Sittel, convinced from reading Hasse about 'the mighty influence of strange blood' upon the nervous system of his patient, encountered some opposition. He then called upon the services of a professor friend, who 'by his moral influence aided me greatly in obtaining the consent of the patient to the transfusion'.[31]

In other situations, consent seems to have been a done affair; very sick patients grasped at this straw of hope. As reported, for example, by the Swedish doctor, Westerberg: 'The patient was informed about the hopelessness of his condition but when lamb blood transfusion was mentioned, the patient eagerly embraced this suggestion and pleaded insistently for it, no matter how uncertain the outcome would be'.[32] In the very few cases reported where a patient refused a blood transfusion (with human or lamb blood), the physician acquiesced.

Medical paternalism, thus, did not imply cruelty or irresponsibility. 19th century physicians based their decisions upon the age-old principle of *beneficence*: a doctor's duty to act in the patient's interest. It was sometimes thought that knowledge might have a beneficial effect on the patient's health. Thus, it could be useful to tell the truth and seek consent.[33] But this was not necessarily *informed* consent. The predominant doctor-patient relationship, Maehle notes, was one where it 'was regarded as inappropriate to expect medical practitioners to educate their patients about the potential side effects of a remedy and to ask them for their consent before prescribing it'.[34] This was a paternalistic attitude, very different from 20th century notions of *patient autonomy*.[35]

But could the doctors realistically inform their patients about what would happen in a transfusion? This is not certain. Many performed a transfusion for the very first time. They only knew from Hasse's reports how they should proceed and what the effects might be. Hence, they may not have anticipated their patients' quite dramatic reactions once the transfusion got started. Many then followed Hasse's advice to continue the operation until the patients claimed they could not breathe anymore. At this stage, one of the Austrian military surgeon Neudörfer's patients tore the cannula out of his veins; this, however, did not stop Neudörfer from performing further lamb blood transfusions and recommending the procedure when no human donor was at hand.[36]

When reports began to appear about unsuccessful cases of lamb blood transfusion, some doctors refused to perform the operation, despite their patients' urgent demands. Boston physician, James R. Chadwick, reported on one such situation with an interesting twist. His account is worth quoting at length:

On one occasion, I was persuaded to go sixty miles to transfuse lamb's blood into the veins of a consumptive. I went after repeated solicitations and a distinct disavowal – on my part – of any belief in the curative agency of transfusion in such diseases. On examining the patient, I found, in addition to extensive disease of both lungs, very labored action of the heart, and obtained the history of much pain and distress in the cardiac region and a number of fainting turns during the previous month. The patient was likewise greatly emaciated. I represented to the man the peculiar danger, which would attend the transfusion of blood into his veins, and finally persuaded him to renounce the project.

A month later, however, a more daring surgeon from New York, a German, successfully transfused six ounces of lamb's blood into the patient. My prognostications of the exceptional risk were fully verified by the unusual symptoms subsequent to the operation. There were 'sharp pains throughout the back, chest and limbs' immediately after the operation. On the next day, again 'acute pains in the back'. On the following morning, 'two fainting spells in quick succession' and a pulse of 130. On the fourth morning, 'palpitation of the heart' for half an hour, and again in the afternoon lasting two hours.

Since that date no untoward symptoms have occurred, but the patient has recently published a card in the local journals announcing that his condition has not been improved by the operation and warning others from trying the experiment.[37]

Was it worth it?

Lamb blood transfusion meant unknown dangers, violent reactions, pain. It was such a new and unknown procedure that it was difficult even to inform about it. Doctors sometimes presented it as an established therapy but most often as an experiment. The aim was to find a new way to cure phthisis, to alleviate pellagra, to counter profuse haemorrhage after childbirth or on the battlefield. And, sometimes, it seemed to have worked.

Blood transfusion was not the first or only experiment then performed on hapless patients. Mental patients, in particular, were often used as human guinea pigs. A report in the early 1880s listed treatments employed in asylums in England at the time: 'hypodermic injections of morphia, the administration of the bromides, chloral hydrate, hyoscyamine, physostigma (the poison from the calabar bean), cannabis indica, amyl nitrate, conium (hemlock), digitalis, ergot, pilocarpine, the application of electricity, the use of the wet pack and the Turkish bath'. In the majority of cases, the drugs merely knocked the patients out for a while but in no way relieved the symptoms. Still, doctors felt that experimenting with one drug at a time might ultimately bring some degree of certainty about what to administer under certain conditions.[38]

Such experimentation at the sick bed was, in the 19th century, not regulated by law but left to medical men, individually and collectively, to deliberate about. As noted by medical historians, there was then 'no precise [historical] moment of moral discovery, no clear or determined march toward ethical imperatives in the practice of experiment', be it on animals or humans.[39] Many 19th century actors possibly agreed with the famous words of Claude Bernard in his *L'introduction à la médicine expérimentale* from 1865:

> So, among the experiments that may be tried on man, those that can only harm are forbidden, those that are innocent are permissible and those that may do good are obligatory.[40]

What category did lamb blood transfusion fall into? Did it do good, was it innocent, or did it only harm the patient?

For many practitioners, the issue was one of *lesser harm*. The physician had to choose between trying an unknown, maybe dangerous but potentially useful, remedy or letting the patient decline and most likely expire. It hurt but it sometimes worked. It did harm but it might do good, and it made future progress possible!

To the Italian alienist Pacchiotti lamb blood transfusion was an easy, safe and successful operation. It was an instance of scientific advance: 'What you today call imprudence is tomorrow the pride of surgery', he argued in reply to his critics and cited as proof some other, previously controversial but later standard, therapies such as the uses of chloroform and ovariotomy. He could have cited (but did not) the 18th century experiments to evaluate the efficacy of citrus fruit in the prevention of scurvy or Edward Jenner's first vaccination trial against smallpox on an eight-year old labourer's son.[41]

A more recent example was that of chloral hydrate, a hypnotic drug introduced in 1869 and used to restrain unruly mental patients. Italian psychiatrists, like Pacchiotti, used it for a variety of indications, albeit with much prudence, given reports of 'chloral poisoning' with serious side-effects – mental irritability, muscular prostration, frequent nausea, and even death.[42] Thus, it was a highly disputed drug. But it was widely used, especially in the Anglo-

Saxon world. Within 18 months of its introduction, around 50 million doses had been dispensed in England alone.[43]

Thus, 19th century experimentation with drugs and treatments was ubiquitous and often drastic. To some doctors, it was their right, indeed their duty, to try out new interventions if they were not obviously harmful.[44] Time would tell if the results would hold. Lamb blood transfusion was an experimental therapy that some physicians, as noted in the conclusions to chapters 4 and 5 above, thought was worth experimenting with until more evidence had been gained. Possible problems would, one Swedish physician assumed in 1874, be ironed out 'while the operation passes through its first year of apprenticeship'.[45] But, as we have seen above, this 'first year of apprenticeship' turned out to be a fairly tumultuous one.

Overstepping boundaries

Lamb blood transfusion was, to many, an irresponsible transgression. It was criticized and ridiculed. When Hasse presented his results at the German Surgical Society's Congress in Berlin in April 1874, the quip went around that it took three sheep to perform a transfusion: the donor, the recipient and the easily fooled doctor![46]

Still, the therapy caught on – and the mockery became more caustic. Panum castigated Hasse and Gesellius as charlatans and lamb blood transfusion as a psychological mystification.[47] In Italy, the celebrated scientist and politician, Paolo Mantegazza, used his contacts with the periodic press to publicly denounce the transfusionists as pre-modern 'alchemists' – daydreamers imagining that the transfused lamb's blood would multiply in the receiving body.[48] His disciple, Enrico Morselli, argued that there was 'something pathological in the psychology of certain enthusiasts for transfusion'. They had willingly let themselves be mystified and misled by the theatrical cleverness with which Hasse and Gesellius had staged the matter, 'helped as it was by the publishing company of the Imperial theatres of Petersburg (!)'[49]

And in Austria, a failed lamb blood transfusion led to a media scandal. It was seen by some as a storm in a glass of water, by others as a serious incident. It concerned a Dr Fieber, an electrotherapist, who at Vienna's General Hospital and with the help of his brother, a surgeon, had tried, and failed, to perform a lamb blood transfusion to a suffering woman, a famous opera singer. A lamb had been procured but would not give any blood and the patient's veins could not be found. The lamb succumbed from the attempts, but the patient was no worse for the incident.

This event, having the ingredients of a celebrity scandal, made quite a stir in the Austrian press. It was called a 'Transfusions-Komödie' and the *Wiener Medizinische Presse* published a satirical 'Chinese Transfusion Story' – a mock letter from a fictitious Dr Tschin, telling of a farcical, tumultuous and failed attempt to perform a lamb blood transfusion in the exotic General Hospital of 'Pecking'.[50] More seriously, the *Wiener Medizinische Wochenschrift* demanded that the hospital administration intervene against Fieber to restore the reputation of the hospital.[51] Other medical journals, however, reacted strongly against this proposal. They found it ridiculous that exaggerated newspaper reports about a trifling incident should lead to disciplinary proceedings; after all, the patient (but not the lamb) had survived. Most of all, it seems, they feared a precedent that would affect the freedom of action of hospital physicians, damage the reputation of the medical profession, and lead to unnecessary disciplinary investigations. As it turned out, no disciplinary action was initiated by the hospital.[52]

Thus, lamb blood transfusion elicited curiosity and controversy, enthusiasm and sarcasm. It is noteworthy that most doctors who tried it, did so only once or twice. A handful performed up to a dozen lamb blood transfusions, and only the Italian alienists and Hasse himself were more ambitious: Manzini and Rodolfi transfused some fifty patients (most of them more than once) and Hasse at least sixty.

By late 1875, reports were largely unfavourable. The attacks and the ridicule in the press, the devastating critique by the physiologists and the discouraging findings of many physicians who had tried transfusion made doctors reluctant to attempt the therapy. It had gradually become de-legitimized. It was now seen as a hazardous play with patients' life and health, an experiment not worth trying. Several physicians regretted having attempted it at all. They swore to never do it again because of the pain and distress that their experiment had caused their patients. 'It is not allowed to endanger a patient's life in order to restore his intellectual faculties', as one Italian alienist warned.[53] And the German doctor Schmidt prophesized that lamb blood transfusion would soon 'like a legion of other remedies invented against [phthisis] fall into oblivion and be counted among the products of human aberration'. His patients, initially reported cured, had soon thereafter died.[54]

After the first year of apprenticeship, thus, doctors and (surviving) patients, Landois concluded, were 'waking up with a heavy head from their initial intoxication with the therapy. The fervour was replaced by a sobering-up.'[55] Or, as another critic expressed it, the initial 'Loblied' for animal blood transfusion was now swiftly turning into its opposite: a 'Schwanengesang'.[56]

PART IV: THE FALL

8. Winding up

On the morning of May 18, 1876, the robber and murderer, Gustav Adolf Eriksson Hjert, lost his head. The event took place on the gallows hill in Vittlånge, a small community in southeast Sweden. Some 200 farmers from the neighbourhood had been commandeered to form a guard with long sticks around the condemned man to prevent him from escaping. At precisely the same moment, at a gallows hill on the isle of Gotland, his partner in crime, Konrad Petterson Lundqvist Tector, was executed. Thus ended the two men's dream to emigrate to America with the spoils of their robbery.[1]

Public executions were popular events in 19th century Sweden – some 3,000 persons had gathered at Vittlånge to hear the criminal draw his last breath and watch the broad-axe fall. What happened next was described by an eyewitness: 'A thick stream of blood gushed forth from the severed vessels and, in that very instant, several persons from the crowd rushed forward provided with glasses and spoons to catch the blood'.[2] The blood's vitality, according to the folklore of the time, could cure a number of severe illnesses: epilepsy, rickets, rabies. That was why the sick jostled at the scaffold with their mugs, pots and rags to gather the headless body's warm stream of blood. It was a not uncommon sight at executions in Sweden and elsewhere in northern Europe.[3]

Here is one account out of several hundred such collected stories from the south of Sweden:

> When a murderer was beheaded, a person could be cured from epilepsy by drinking three tablespoons of the executed person's blood and then walk backwards from the scene, as many steps as possible, since that many years you would be free from the disease [...] Normally you tried to walk 100 steps backwards to be certain of being healthy for the rest of your life.[4]

Sometimes, the authorities gave permission to approach the headless body; at other times, the bailiff drew his sabre to prevent the sick or their helpers from coming too close to the scene.

Not any kind of blood was considered effective, however. It had to come from an executed criminal, from a soldier who had died in battle, from someone who had murdered but escaped justice, or from someone who had met with a sudden death. Such individuals possessed extraordinary powers or had experienced extraordinary events. They were outside the normal; therefore, their blood had a special and healing force. To drink it, would break the curse thought to lie behind, in particular, the epileptic fits with their frightening attacks of delirium, cramps and unconsciousness. With their blood, the evil or the violently dead, would give the innocent sick their life back.[5]

**

These two executions were the last ones in public in Sweden. Three years later, public executions were banned. Further executions would, until the punishment was abandoned in 1921, take place in the seclusion of prisons; the last execution was performed in 1910. After 1876, thus, sick people could no longer drink the presumably life-giving blood from a just-beheaded murderer.

The year 1876 was also one when virtually all animal blood transfusions disappeared from the scene. This procedure, too, had been seen by many as something almost magical. Transfusions were 'strange experiments' that sometimes made the almost-dead rise up as through a miracle.[6] The opening of the artery of a lamb to let its blood pour into a patient may have been done with medical, not mystic, arguments and taken place in a quiet clinic, not in the turmoil of a gallows hill. It used syringes, cannulas and rubber tubes, not the broad-axe or the mugs and spoons of poor people. But the end result of a lamb blood transfusion was often as illusory as that of the drinking of a murderer's blood. Transfusions were strange experiments that often failed; death could not be averted.

Now, this procedure, too, would soon be history. No further lamb blood transfusions were reported from Germany or Austria, and only a few additional attempts were made in Italy to cure the mentally ill with the blood of a lamb. The last one may have been done by Dr Ernesto Dallera in Genoa; he reported having made a lamb blood transfusion in early January 1876 to a mentally disturbed woman.[7] In 1878, there were two direct lamb blood transfusions in Sweden and three on a battlefield in Bulgaria, and in 1879 there was one in Iowa, to a woman suffering from consumption.[8] Thereafter, nothing, it seems. Animal blood transfusion disappeared from the arsenal of therapy about as abruptly as it had appeared a few years earlier.

So, too, did soon virtually all transfusions with *human* blood. This is in many ways more surprising. How could that happen?

The condemnation

On August 2, 1883, Ernst von Bergmann, the renowned military surgeon and professor of surgery, gave a celebratory lecture at the Berlin Academy of Military Medicine. He had chosen as his subject, *Die Schicksale der Transfusion im letzten Decennium*. 'The fate of transfusion in the last decade' was obviously something that engaged both him, the military and the medical audience assembled. The lecture was later referred to as the nail in the coffin of 19th century blood transfusion.

Transfusion, von Bergmann told his audience, was an operation that, until very recently, 'had been hailed as the most significant and important of modern surgery, one that would inaugurate a new era within the whole of medicine and that more than any other method had seemed capable to sustain the receding life of the wounded'.[9] Animal blood transfusion, especially, had received a triumphal reception in the hospitals and was predicted to have a glorious future within surgery. But now, the transfusion instruments collected dust, the exalted expectations were abandoned, the enthusiasm had waned.

To von Bergmann, it was particularly difficult to understand why anyone could think that animal blood transfusion – so terrifying in its effects! – was worth trying at all. Certainly, some attempts had seemed successful – at least the patients had not died! – but such clinical successes were worthless, if they did not rest upon a solid foundation of physiological knowledge. In this, as in other medical matters, science must have the last word.[10]

Von Bergmann's physiological arguments against lamb blood transfusion had been heard before, from Panum, Ponfick, Landois and others; we learnt about them in a previous chapter. But von Bergmann went one step further. Based on research done 'in the last decade', he dismissed transfusion also with defibrinated *human* blood. It destroyed the recipient's blood cells and brought about a shock to the organism. The only transfusion worth doing was one that moved blood directly from the artery of a human being into the vein of a needy patient – but then, von Bergmann added, it was uncertain whether 'an intervention that requires so much sacrifice from a fellow human being will ever come into general use'.[11]

Von Bergmann could have quoted (but did not) the French professor of clinical medicine, Georges Hayem, a pioneer in the field of haematology. He had done extensive research on the nature of blood, and as recently as the year before – 1882 – published a more than 500-page volume about the effect on blood of various medications and therapies, including transfusion. Hayem had arrived at much the same conclusions as von Bergmann gave vent to. Animal experiments and clinical experience had convincingly shown that transfusion with animal blood was terrible. But the use of defibrinated human blood was almost as harmful. In both cases, reactions were too unpleasant, even lethal, and indications too uncertain for a transfusion to be of any clinical use. Hayem's conclusions were as categorical as those of von Bergmann: transfusion could be recommended only in cases of severe haemorrhage menacing the very existence of the patient. But since the physician then must use whole, and not defibrinated, human blood, the operation was so unwieldy that it was scarcely worth attempting at all:

> How could we ever, on the battlefield, in an accident or even in a surgical ward manage to do a whole blood transfusion to an individual suddenly close to death, when in a laboratory, where everything is organized for the operation, some of our dogs have succumbed before our very eyes before we had the time to perform a transfusion?[12]

The fate of transfusion, with human or animal blood, seemed sealed.[13]

The medical needs, however, were still there. This raises the question of what the alternatives were. To approach this issue, we must make a detour. Why do a blood transfusion at all? What exactly was the rationale for moving the blood of others into sick people? This had been a matter of contention ever since Blundell re-introduced transfusion in the early 19th century. Blood was indeed 'a very special fluid', as Goethe let Mephistopheles say in *Faust* – but what kind of fluid was it and what did it do in the body?

Understanding blood

In the beginning of the 19th century, an old *vitalistic* notion still lingered stemming from Hippocrates. Blood was one of the four central humours, perhaps the most important one. By then, the idea that a person's personality expressed itself through the blood may have disappeared in medical circles; still, many thought that blood in itself had a life-giving capacity. A transfusion would transfer vitality in an almost magical way. This notion was expressed in the rhetoric of 'reawakening' lifeless experimental dogs and 'reviving' patients with transfusions of blood. It could be heard even in the mid-19th century. Accounts of transfusion then regularly mentioned that the donor (if it was not a sturdy lamb) was a 'robust young man', a 'flourishing young women', 'two strong seamen' and the like. This was most likely meant to underline that donating blood was not for the weak but could also be interpreted as an implicit wish for the donor's strength and vitality to be moved into the waning patient.

Over the years, a *biological*, or functional, view of blood took over.[14] The blood's recognized components were seen as separate biological entities; especially the red blood cells were essential. They had the biological function to transport oxygen from the lungs to all parts of the body and waste products back to the lungs. The role of a blood transfusion would then be to replace the missing red blood cells in patients suffering, for example, from anaemia or intoxication. It might even, some physicians argued, stimulate the production of new blood cells in the recipient. Italian psychiatrists, for example, maintained that since mental patients had a deficiency of red blood cells, a blood transfusion, even one performed with the blood of a lamb, would have a both stimulating and nutritious effect. Hasse initially saw the transfused animal blood as a curative as well as a palliative drug for a variety of afflictions; later he changed his mind to consider it mainly as a nutritious agent.

As to the other components of blood (the fibrin, the white blood cells, the platelets) physiologists were uncertain and sometimes disagreed strongly about their function in the body. Most hotly debated was fibrin. To lamb blood enthusiasts, defibrinated human blood, coming from a vein and subsequently 'beaten to death', was terrible – an argument resonating with vitalistic overtones. Physiologists took a more biological view. In the 1830s, the French physiologist, Magendie, had found that fibrin was central for blood circulation since it helped the blood pass into the capillary system. His results were contested by Panum and others, favouring defibrinated blood to avoid transfusing dangerous blood clots into the recipient, but were re-instated again as fundamentally correct by, for example, Hayem.[15]

Thus, the very same blood component could be seen as either nutritive or destructive, just as animal blood could be considered as either useful or dangerous. Many medics agreed, however, that for blood to be harmless it should be transfused slowly and in only small amounts at a time, in line with how strychnine and morphine were administered when used as remedies.[16]

In the 1880s, a more *mechanical* view of blood came to influence medical thinking and therapies.[17] Chock and anaemia were now diagnosed as the result of there not being enough liquid in the vessels. The patient's blood pressure was insufficient to keep the body's machinery going; the blood became stagnant and did not circulate at a sufficient pace. Since blood was now considered a mere transport medium, it could be replaced by other substances to keep up the volume and pressure of the vessels and get the circulation working again. So, why not try milk, Gum Arabic or a saline solution? Milk was used in the US; Gum Arabic, here and there, and saline solutions had, with some success, been tried in cholera epidemics since the 1830s.

Figure 27. Charles E. Jennings' combination of transfusion of blood and a saline solution (Jennings 1896, 331).

Extensive animal studies from the 1880s onwards supported the utility of various varieties of saline solution and practicing physicians eagerly embraced this new therapy against shock, anaemia and other afflictions. A saline solution was easier to obtain and less risky than blood; it did not coagulate and did not require troublesome surgical incisions in the donor's body (since no donor was used). By the turn of the century, saline infusions prevailed, and blood transfusion was, as one historian phrased it, 'relegated to the quaint pages of medical history'.[18]

Thus, ended the 19th century story of blood transfusion. Occasional and isolated transfusions would still be performed in desperate situations. In the US there was even, in 1890, an allegedly successful transfusion with blood from a lamb.[19] But blood transfusion had not become part of standard hospital interventions; it was considered useless, cumbersome and, often, dangerous. It was seldom mentioned in medical handbooks and professional journals, and lamb blood transfusion only as an anomaly.

Lessons learnt

It is time to sum up the lessons learnt from the short and confused episode of lamb blood transfusion. So much hope had been invested in this therapy by physicians and patients alike, so much scorn heaped upon its use. Perhaps the story is typical of what happens when a new intervention is proposed within medicine: there is an often tortuous process before it is accepted or dismissed as a standard therapy, with successful trials and failed attempts.

The lamb blood experiment was unique, however, in its scope and intensity. It was no local German affair but a novelty that spread with surprising speed across the European continent and to Scandinavia, England, the USA, even Chile. It arose great public interest, was suggested by military surgeons and tried out by renowned physicians and psychiatrists. Its benefits and drawbacks led to heated discussions at meetings and in medical journals. For a while, it seemed to promise salvation for such terrible afflictions as phthisis and pellagra. Then, it was dismissed as useless by physiologists and disappointed doctors alike.

Transfusion in general was an experimental therapy during the 19th century, difficult to perform and based on incomplete physiological knowledge. Indications varied, so did methods and techniques. This theoretical uncertainty paved the way for physiological and clinical experimentation with defibrinated blood, arterial transfusion, capillary transfusion – and for the reintroduction of lamb blood transfusion.

For the German medical historian, Barbara Elkeles, the central question concerning lamb blood transfusion was the naivety of doctors. Why were they so enthusiastic about a therapy that was so painful for their patients? She does not answer this question but notes the scepticism and reluctance among many practicing physicians.[20] However, and as we have seen in this book, this is not the whole story. Many doctors claimed success. They saw their phthisis patients improve, their mental patients recover their speech and appetite and their anaemic patients gain a new strength. They reported their good results and other doctors followed suit. Perhaps very little blood was transfused in each transfusion – something that may account for the patients surviving the intervention. Perhaps they would have recovered anyway, given the food and care they received in hospitals and asylums. Still, for desperate patients and physicians, and at least for some time, lamb blood transfusion seemed a beneficial, albeit often painful, last recourse – and sometimes, it worked!

The experience of lamb blood transfusion also highlights the difficulty to draw a line between, on the one hand, cutting edge therapies based on theory and animal experiments, and audacious sickbed experimentation, on the other. It is not easy today, as Elkeles notes, but it was, as I have discussed in this book, even trickier in the 1870s. The results of animal experiments were not automatically relevant for sick human beings, the physiological nature of blood and its function in the body were still largely unknown and clinical experience, too, was contradictory and incomplete. When the good results became fewer and fewer, and initial patient recovery turned into *status quo ante*, transfusion was abandoned in favour of other, less contentious, fluids than blood.

Still, the account of the rise and fall of 19th century blood transfusion cannot end with its condemnation and medical abandonment. It is indeed a story of how a contested medical therapy was used and argued about, and its social circumstances. But it is also, and importantly, a story of human ambitions, emotions and ingenuity. We have got to know some central actors involved in the struggles for or against lamb blood transfusion. Before ending, I therefore want to reconnect with the main protagonists of this story. How did they react to the denunciation and disappearance of transfusion from the therapeutic arsenal? What happened to them afterwards?

Human trajectories

By the late 19th century, the dust had settled on the lamb blood controversy. Most of those involved were either dead or had left the transfusion scene. Panum died in Copenhagen in 1885 at the age of 65. Landois, too, was 65 when he died in Greifswald in 1902. Von Bergmann lived on until 1907, Ponfick until 1913, Hayem until 1933. These men had been prominent within their respective scientific fields, had published widely and received honours of various kinds.

Of the Italian lamb blood transfusionists, Livi died in 1876, Ponza in 1879 and Manzini would soon use other means of redress for his patients in the Brescia asylum. His colleague, Rudolfo Rudolfi, got involved in various public health initiatives, especially for the care of poor children, and Lombroso became the founder of the controversial Italian school of anthropological criminology. Manzini and Rodolfo died in the 1890s, Lombroso in 1909.

This leaves us with our three central characters: the still unknown until the early 1870s but from then on internationally famous physicians, Oscar Hasse, Franz Gesellius and Joseph-Antoine Roussel. Their lives after the crucial year of 1874 took quite different paths.

By 1875, Hasse had made some 60 lamb blood transfusions. He hoped to make 200 in order to publish a more complete account of the therapy.[21] Of this no more was heard. After the attacks by Landois and Panum, Hasse returned to a more anonymous life as a local doctor in Nordhausen. There, he would become famous for saving lives through tracheal surgery, an operating skill learnt in Berlin in the early 1860s.[22] He initiated a popular hiking club for walks in the nearby Harz Mountains and became its first president.

Hasse died in 1898. He is today remembered as a prominent son of the city. There is a street named after him and an imposing memorial in the Nordhausen town park.

As for Gesellius, he abandoned transfusion altogether after his debacle at the St. Petersburg competition in 1874; his transfusion (with lamb blood) had been messy and painful, and the patient died. He left the medical field in 1875 to start a German language newspaper, *St. Petersburger Herold*, and be its chief editor. Based on somewhat questionable journalistic methods, it was a success, and Gesellius became a well-known, though disputed, figure in the city's social life. In the 1890s, he had to abandon newspaper ownership due to financial problems. In 1914, the paper ceased publication because of the war. By this time, Gesellius was long-since dead; he passed away in 1900.[23]

Figure 28. The Oscar Hasse-Medallion on the Hasse-Gedenkstein in Nordhausen (Photo: the author, May 2019).

And Joseph-Antoine Roussel? His fate and that of his ingenious transfusion apparatus were closely tied to military demands. This situation merits a somewhat longer account. It was, after all, military surgeons who had first entertained the idea of animal blood transfusion and who had eagerly defended it in the early and mid-1870s.

We last met Roussel in 1874/75 when he made energetic tours across Europe to demonstrate his apparatus both in civilian hospitals and to various armed forces. He was quite successful in this endeavour despite acrimonious comments from competitors.[24] By now, he had influential and highly placed allies in several countries. In January 1874, the Austrian military surgeon, Neudörfer, endorsed the Roussel *transfuseur direct* to the Austrian war ministry and soon thereafter, Roussel's collaborator Heyfelder recommended it to the Russian War Ministry. Here, too, it was adopted and a large number of transfusors were reportedly ordered and paid for by the Russian government. As an added bonus, Roussel received the prestigious, Order of Saint Vladimir.[25] He then demonstrated his apparatus in Belgium whereafter the Belgian army adopted it in July 1876 and Roussel was appointed, Chevalier de l'Ordre de Léopold.[26]

Roussel's good fortunes continued across the channel. He may have lamented in 1877, 'I hope that the English Government will not be the only one which in case of war would allow their wounded soldiers to die of haemorrhage from the want of their surgeons being instructed in the practice of transfusion'.[27] But soon thereafter, and on the recommendation of several leading British surgeons, his apparatus was introduced into the ambulances of the British army and marine.[28] No real cause for complaint, thus.

The French Ministry of War was more difficult to persuade. It had waited until after the 1870/71 War to adopt a French instrument, the Colin transfusor.[29] In 1879, due to complaints about its suitability, it was deleted from the reglementary register. By then, a committee consisting of, among others, Claude Bernard had recommended the Roussel apparatus as the best alternative but it was still not acquired by the French authorities.[30] In 1881, the question came up again and Roussel complained: despite petitions from esteemed physicians and surgeons (this time including Hayem), despite the approval given by the commission in charge of the selection of surgical instruments for public authorities as well as by the *Conseil de santé militaire*, 'not a single *transfuseur direct* had yet been acquired for the ambulances and the French hospitals'. But, he added, 'I have fought for fourteen years, still I am not discouraged'.[31] Some years later, Roussel's apparatus was part of the French military supply.[32]

By this time, Roussel, seeing the decline in transfusion interest, had abandoned it for a new therapy – hypodermic injections. It could be restorative or calming or used as a purgative. In the 1880s, he experimented with himself, injecting iron, arsenic, mercury, phosphates, eucalyptol, menthol, even phosphor; he also tried out various dissolvents. Hypodermic injections could, he argued, be used for a variety of indications.[33] He invented a new type of syringe for the purpose and in 1888 started a journal, *La Médicine Hypodermique*. It was published with, what seems, some success for about ten years. Roussel died in 1901.[34]

Figure 29. Blood transfusion in the French army using Roussel's instrument (Delorme 1888, 529).

No more blood on the battlefield?

One large, belligerent nation did not adopt the Roussel transfusor – Germany. This is somewhat surprising given the 'great, widespread enthusiasm for the transfusion of blood' in its military circles after the Franco-Prussian War. But it waned quickly due to condemnations by von Bergmann and others. The lamb blood alternative was relegated to the realm of the improbable and so was transfusion of human blood. Just as in the 1860s, military surgeons thought it difficult to get hold of human blood in war conditions: donors had to be healthy, rested and strong. Even the newly invented methods of injection or infusion with a saline solution seemed to them unworkable on the battlefield. Salt and hot water were difficult to get hold of there, and how would a surgeon find the necessary time to infuse the liquid as slowly as was deemed necessary?[35]

The introduction of antiseptic and aseptic procedures also meant that secondary or late haemorrhages, upon which almost all transfusions had been made during the 1870/71 war, were now less likely to occur. A transfusion would therefore not be called for.[36] For acute interventions, some military surgeons suggested a new remedy – autotransfusion. The legs of a wounded soldier could, already on the battlefield, be tightly wound by elastic ribbons. This manoeuvre would concentrate the blood in the rest of the circulatory system and thereby keep up the pressure and give the heart enough blood to work with until a transfusion with a saline solution could be performed in the field hospital.[37]

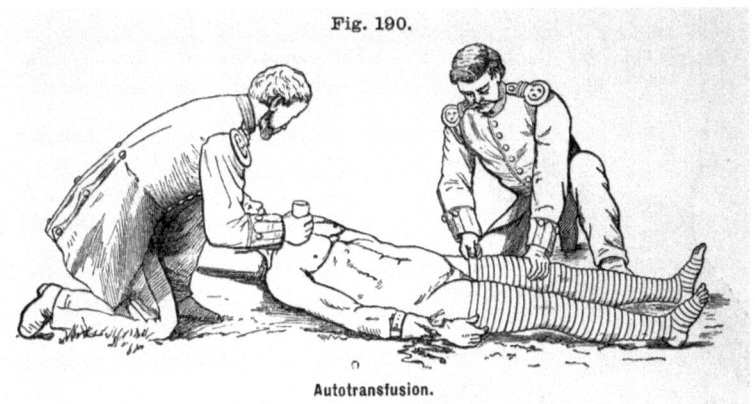

Figure 30. Autotransfusion on the battlefield (von Esmarch 1894, 117).

***

A final word, for now, about the diminishing need for transfusion on the battlefield, using a citation from the German doctor Friedrich Wilhelm Hertzberg. Referring in 1869 to the butchery of the 1866 Franco-Austrian War and the impossibility of providing all the wounded with new blood, he laconically stated:

If 'you really wish to eliminate the bad effects of bloodshed in war, well, then abolish war'.[38]

After the Franco-Prussian War and the Russian-Turkish War of 1878, this prophecy indeed seemed to have become true, at least in Europe. There were no more major wars on the European continent until 1914. Then, as we know, blood inundated the battlefields. Once again, blood transfusion returned, this time to stay. It now came in the form of indirect transfusion using bottles where the donated blood had been mixed with a citrate solution to prevent its coagulation. The procedure was introduced by Canadian, then British and American physicians, and thus *not* by the German or Austrian military surgeons whose predecessors had been such ardent advocates of blood transfusion some forty years before.[39]

Epilogue: The return

I am on my way to the Paris Museum of Medicine lodged in a corner of the huge École de Médicine in central Paris. I want to look at the Roussel *transfuseur direct* exhibited there. Halfway up the winding staircase to the museum I have to stop and look. Tucked away in this obscure place is a large painting showing a dramatic medical intervention. It's subject? *'Transfusion du sang du Chèvre'*.

The painting's central figure is a bearded man, presumably a doctor, surveying the transfusion of goat's blood to a seemingly unconscious young woman. He is assisted by two men in butchers' aprons and two colleagues in black suits. One of them introduces a cannula into the patient's vein, the other keeps check of time. In the background, a nurse is busy arranging test tubes and other medical paraphernalia. She has turned away from the scene; perhaps she disapproves of this attempt to move animal blood into a helpless patient?

The painting is by the young French artist, Jules Adler, best known for his realist depictions of common folk. Perhaps he felt that the doctor commissioning the painting was on the side of the working classes. It was the Paris physician, Samuel Bernheim, a tuberculosis specialist who had established a charity to send poor patients and their children to the seaside as part of the sanatorium movement. The painting was exhibited at the Paris Salon of 1892 where it was well received and won an award.

The painting is intriguing. Was animal blood transfusion being reintroduced in France, a country that twenty years earlier had been completely disinterested in the therapy? The answer is yes, but only for a short while and with a very different physiological rationale than before. Medicine had changed after Koch's discovery of the tuberculin bacillus in 1882 and Pasteur's immunological research. Blood was once again seen in a different light.

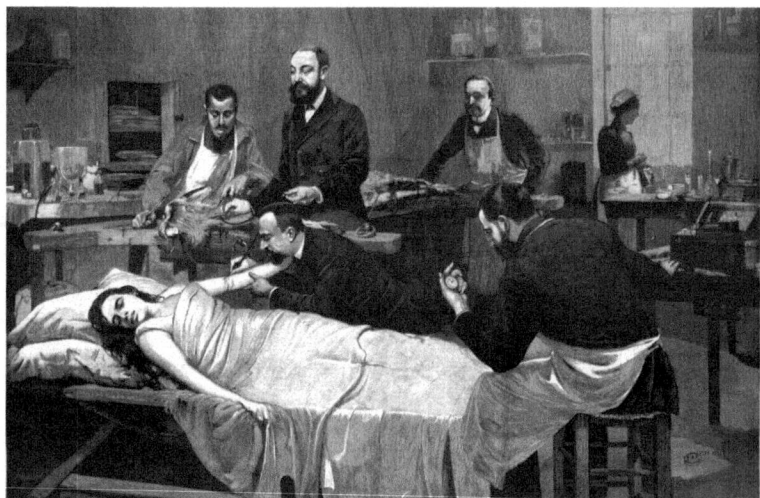

Figure 31. Jules Adler, Transfusion du sang du chèvre. Engraving by Henri Meyer (Le Journal Illustré, May 22, 1892).

The advent of serotherapy

Here is Bernheim describing his work in a lecture to the Société des Practiciens de France, published in *Le Moniteur Médical* in March 1891. He had, he reported, 'in the last two months' made thirty-three transfusions with goat's blood to tuberculosis patients, and with astonishing success. One of them may have been the young woman in the painting:

> Miss B., nineteen years old, residing at 4, Boulevard du Temple, lost her father to tuberculosis. Three of her brothers died from the same disease. She herself, ill since six months, has tuberculosis to the second degree on her upper left lobe. Three months ago, Koch's bacillus was detected in her sputum. The patient has been treated with two transfusions at an interval of fourteen days.
>
> Today, the patient no longer coughs, the expectorations have disappeared. The young girl has been greatly strengthened; she eats and sleeps well. We can no longer discover any trace of lesions on her left top lobe and she breathes normally [...]. No more bacilli in her sputum.[1]

At about the same time, between the end of December 1890 and March 1891, two colleagues in Nantes, Georges Bertin and Jules Picq, made subcutaneous injections with goat's blood into some fifty patients. The procedure was repeated every fortnight, each time with about 15 grams of blood and with positive results.

These injections and Bernheim's transfusions were based on a different physiological reasoning than what we have encountered earlier in this book (though Bernheim was quite vague about why his transfusions worked). The physicians were not interested in moving oxygen-rich red blood cells into the patient's organism, neither did they intend to fill up the vessels to prevent a loss of blood pressure. Instead, they seemed to see the transfused or injected goat's blood as a *biochemical* substance. The aim was to transfer the animals' innate natural immunity to tuberculosis as a kind of vaccine to the suffering patients and thereby help them resist the dreaded disease.[2]

These were sensational ideas. Bernheim went on to treat some ninety phthisis patients, all of whom, he reported, asked for a second transfusion after having happily experienced the first one. The procedure was now being studied by a professor at the Faculté de Médicine.[3] Bernstein made public demonstrations, gave interviews and had the procedure depicted in newspaper images as well as commanding the painting by Adler. His and the Nantes group's successes were reported as far away as Australia, New Zealand and the USA.[4]

Serotherapy was, indeed, a newsworthy subject. Only a few days before Bertin's and Picq's injection experiments, Behring and Kitasato in Berlin had published a seminal study on diphtheria and tetanus immunity; Roux and Yersin in Paris were also on the track. Based on these studies, serotherapy for large-scale treatment of diphtheria would soon be undertaken in both Germany and France. The technique was to induce immunity in host animals, normally horses, and then bleed them, separate out the serum and inject it into humans. In the mid-1890s, this represented a major therapeutic innovation and an important element of public health policy in France as well as in other European countries.[5]

Less publicized were the attempts by the Italian professor of pathology, de Domenicis, who in 1894 repeatedly injected dog's blood into patients suffering from anaemia and tuberculosis. He, however, only obtained limited results and only in a couple of not very sick patients.[6] Still, the French serotherapy researcher, Charles Richet (a future Nobel Prize winner), was positive about the attempt and argued that dog-serum could improve the patients' general condition by its remarkable stimulating properties. Patients would thereby improve their capacity to resist the infection.[7] As it turned out, however, neither the French goat nor the Italian dog blood therapy was a long-term success; both practices were soon abandoned.

Hasse vindicated?

Then, a few years later, in 1901, animal blood transfusion was again on the agenda. This time, the advocate was none other than August Bier, professor in Greifswald, later the successor to von Bergmann in Berlin and one of the most prominent German surgeons of the early 20th century. And again, new physiological notions were advanced to underpin the use of this therapy.

Bier had made animal experiments with surprising results and found interesting similarities to how Hasse's patients had been affected by lamb blood transfusion back in the 1870s. In an article in the *Münchener Medizinische Wochenschrift*, Bier now forwarded the idea that 'alien' blood could work in the same way as a mild infection: it would kill or weaken certain bacteria in the sick body, plus stimulate digestion and appetite. Based on this idea, he injected over a period of three months (November 1900 to February 1901) defibrinated lamb's blood in small doses (thirteen times in all) into the veins of a young man suffering from severe tuberculosis. At the time of publication, the patient was feeling much better, had an appetite and moved about. Ten other, almost dead, tuberculosis patients were treated in the same way. Three of them had died but the others had gained weight and appetite, and two had already left the hospital. These results convinced Bier that blood from a lamb could work as a remedy if transfused in his suggested careful way.[8]

After this isolated announcement, no more was heard from Bier about the subject for another twenty years. But in 1921, he returned with an extended analysis of why lamb blood transfusion was useful against inflammatory diseases, like tuberculosis. Again, it was the patients' substantial improvement in weight and wellbeing, appetite and blood condition that inspired him. And again, Bier saw Hasse as somewhat of a pioneer who should be given the respect denied him in the 1870s. To Bier, Hasse had been unjustly attacked, ostracized and treated as a swindler. The result was that 'the transfusion of strange blood was buried by Landois and Panum. The funeral oration was given by v. Bergmann who even condemned transfusion of species-similar blood into earth and ground'.[9] Still, Bier argued, Hasse had been right in transfusing lamb's blood, though initially for the wrong reasons, and he should really not have given such excessive dosages of blood. Bier instead, just like the Italian alienists in the 1870s, transfused only small doses of blood, but repeatedly, and to good results.

Bier now thought that a lamb blood transfusion could act as a benign infection that, if administered correctly, would be valuable to the organism. He saw his targeted injections of animal blood as an example of 'protein-body therapy', a homeopathic treatment coming into fashion in the early 1920s. A blood transfusion should not serve to replace blood, as had been believed for two hundred years. Instead, Bier argued, the 'foreign' blood would act as a trigger on the organism, which had to react by mobilizing its defences. The disintegration of the blood cells – what Landois and Ponfick had seen as a lethal effect of animal blood transfusion – would, Bier claimed, work as a stimulus on the inflammatory processes in the sick body. So, too, would the dreaded side-effects: the shortness of breath, the dry cough, the red and hot skin, the increased peristalsis, and fever. They would aggravate and thus heal the inflammation.[10]

This seems to have been Bier's last words about animal blood transfusion, but he inspired others. In the early 1920s, von Klebelsberg, a local doctor in the Austrian town, Hall in Tirol, injected repeated small doses of defibrinated blood from recently slaughtered cattle into seventy-two mental patients. His idea, similar to Bier's, was to use blood as a stimulus to the organism, somewhat like injections with cocaine or milk. The reactions were in some cases very strong: several, already quite sick, patients died; some had panic attacks while others did improve after each injection but only for a while.[11]

Later in the 1920s and early 1930s, Bier's colleagues in Berlin, Zimmer and von Balden made intravenous injections of lamb and ox blood in six cases of Basedows disease, reportedly with success.[12] Hadenfeldt made 200 intravenous injections of animal blood, mainly into cancer patients with, as he argued, good results. He used blood from different animal species (calf, pig, lamb, even horse) but the blood of lamb proved to be the most suitable alternative and also the easiest to procure. Hadenfeldt's conclusion, following Bier, was that this procedure was 'a method – stronger than any other that I know of – to give a certain respite and force an improvement in the general condition of the patient and thereby, in many cases, make an operation possible at all'.[13]

Soon, however, there were no more reports of such attempts based on protein-body theory. The idea that species-alien blood would destroy and disintegrate the recipient's blood cells, and thereby stimulate the body's defences and encourage the formation of new healthy blood and tissue thus seemed to have been abandoned, for this time and perhaps for all.

French interventions

Meanwhile in France, there were a few more attempts. They were based on more conventional notions of the value of an animal blood transfusion.

In 1916, a private practitioner, Dr Famius, writing in the popular journal, *La Science et la Vie*, described his many transfusions 'during the last forty years'. Some of them had been made with goat's blood when no human donor was at hand. Famius preferred goats to other animals since their blood cells were much smaller than human ones and because they were immune to a number of serious diseases. His favourite was the 'Lamartine' type of goat: small, with long white and silken fur, hanging udders, disappearing horns, a soft skin and a mild gaze.[14]

Famius' efforts were not well known. Ten years later, René Cruchet, professor of pathology and general therapeutics at the University of Bordeaux, would make a greater stir, also internationally, with his animal blood transfusions. Cruchet saw himself as the successor to Jean-Cyprien Oré, a lamb blood proponent whom we have met in previous chapters. Not only was Cruchet at the same university as Oré had been in the 1870s. He also wanted to return to Oré's 'classical studies' showing that animal transfusion was superior to the man-to-man alternative. Thus, Cruchet had, 'on a more modern basis', made numerous animal-to-animal experiments in the laboratory as well as injected sheep's blood and transfused horse's blood into tuberculosis sufferers and mental patients. The trick, Cruchet argued, was to introduce the blood very, very slowly, especially in the beginning of the two-to-four minute operation and to dilute the blood with physiological serum. One of his patients had died, two other remained unchanged, but the state of several patients had improved, 'and they have even asked urgently for repeated transfusion', he reported in 1926.[15]

Again, however, the use of animal blood transfusion did not meet with universal acclaim, far from it, and it would soon disappear. An anonymous writer in the *British Journal of Surgery*, reviewing Cruchet and his colleagues' 1928 volume on transfusion, was outright sarcastic. I will finish my book with this verdict:

The reader's own blood runs cold as he reads the account of how severely the patients suffered from the expected symptoms and how narrowly they escaped death, and he is not consoled by the preliminary statement that, 'le mouton choisi était superbe'. No evidence is forthcoming that any improvement that the patients may have experienced was due to the treatment [...]

The authors nevertheless suggest in conclusion that the transfusion of blood from animals to man may soon fulfil a 'rôle énorme' in the treatment of disease. They have visions of stables with horses and of 'moutons superbes' immunized against every form of bacterial infection, including filter-passing organisms, with rivers of blood pouring from them into the veins of suffering humanity. We fear that they are, to say the least of it, sanguine.[16]

***

Still, one cannot help but wonder what happened to the patients who were among the first to undergo the trial of a lamb blood transfusion. They were sick and desperate, and they were brave. What became of the thirteen-year old girl, Hermine Krüger in Schwenda? Or Carl Jacobsson, coughing his lungs out in Mösseberg, or Annunciata Rossi, one of the emaciated and depressed peasants in the San Lazzaro asylum in Reggio Emilia? Initial reports told of their improvement from 'the mighty influence of strange blood'. But how did they fare thereafter?

The archives finally produced some answers. Annunciata Rossi did not recover, after all. She died soon after the transfusion, at the age of thirty-six. She left a husband and a daughter; two other children had previously died.[17] Hermine Krüger, the very first person treated by Hasse with a lamb blood transfusion, was luckier. She recovered, married a man from Schwenda, had a daughter and lived on until 1903. Then she died of lung complications at the age of fifty-three.[18]

And Carl Jacobsson? He was the young man, severely ill with a lung affliction, who became the first patient in Sweden to receive a lamb blood transfusion. We last heard of him in 1875, one year after the intervention; he seemed to have recovered well. He then worked as a labourer, married a much younger woman in 1889, had no children and spent the rest of his life in the same village in northern Västergötland. There he died in 1929 at the ripe old age of eighty-one.[19]

Notes

Prologue

1 Hasse 1874a, 15–20.

Introduction: 'The mighty influence of strange blood'

1 Landois 1875, iv.
2 Björnström 1873/74, 180.
3 Berner 2012.
4 Guerrini 2003, x-xi.
5 Sittel 1874, 158.
6 The book is based on material in six languages and stays close to the actors' accounts. All translations, unless otherwise stated, are by me. The quotations are sometimes slightly edited for readability and long paragraphs are broken up.

1. Using the blood of others

1 Sander 1874, 173.
2 Köhler 1906, 293.
3 For a history of transfusion, see Starr 1998.
4 Gilder 1954, 172.
5 Marinozzi et al. 2018, 2.
6 Scheel 1802, 232–233; Maluf 1954, 68.
7 Quoted in Zimmerman & Howell 1932, 419. The contest for priority between England and France and the dramatic events in Paris are described in Moore 2003 and Tucker 2011.

8   Riesman 1937, 191.
9   Marinozzi et al. 2018, 3.
10 Scheel 1802; Dieffenbach 1828.
11 E. g. Martin 1859; Oré 1868.
12 – (1834), 157.
13 Pelis 1999.
14 Pelis 1999, 15.
15 Pelis 2001a, 206.
16 Marinozzi et al. 2018, 5.
17 Pelis 1997, 336–337.
18 Panum 1864.
19 For a summary of the arguments, see Schorr 1956, 32–36.
20 Roussel 1885, 44.
21 Casse 1874, 3.
22 Leisrink 1872, 237.
23 Billroth 1875, 3.
24 Starr 1998.
25 Greenwalt 1997, 555.
26 Calculated from the overviews in Martin 1859; Oré 1868; Marmonnier 1869; Sacklén 1870; Landois 1875.
27 Marmonnier 1869, 35.
28 Severn 1839, 249.
29 Schorr 1956, 38–43.
30 *Daily Alta California*, July 2, 1871.
31 Reported in *Transactions of the Medical society of North Carolina* 1872, 40–45.
32 Winants 1872, 110.
33 *The Obstetrical Journal of Great Britain and Ireland* 1873, 360; also Albini 1872.
34 Hasse 1874a.

2. Ambitions and connections

1   Jürgensen 1885, 274.
2   https://www.google.com/search?client=firefox-b-d&q=sanguine+definition.
3   Müller 1955, 203.
4   Hasse 1868.
5   Hasse 1869; Hasse 1874a.
6   Gesellius 1873.

7   Hasse 1875, 252.
8   [Seyfer]th 1874, 46.
9   Hasse 1874b, 408.
10 E.g. Penzoldt 1874, 502.
11 See https://www.nordhausen.de/allgemein/cblock_lang.php?CBlNr= 11508.
12 Grosberg 1926, 10.
13 See e.g. Leisrink 1873, 265–272; Panum 1875a, 47–91.
14 Daniel 2013.
15 Hasse's book reports on sixteen transfusion with human blood and fifteen with lamb blood; thus, the title referring to 31 lamb blood transfusions is misleading.
16 Gesellius 1874, 17.
17 The university archives of Greifswald university contain information of both Hasse's and Gesellius' student pranks, something which in Gesellius' case, but not in Hasse's, led to him being asked to leave the university.
18 Hasse 1875, 253–254.
19 Roussel 1875a, 141.
20 Roussel 1885, 22–24.
21 Heyfelder 1874 in Roussel 1876a, 160–163.
22 Buess 1953, 253.
23 Olivier 1952, 9.
24 Full accounts in Roussel 1875a, 136–139; Roussel 1877, 54–56.
25 Roussel 1875a, 139; Roussel 1877, 56.
26 Roussel 1885, 72.
27 Roussel 1884, 812.
28 Roussel 1877, 38.
29 Roussel 1885, 51.
30 Braun 1878, 89.
31 Roussel 1875a, 151.
32 Landois 1875, 324–325.
33 Roussel 1877, pp. 42–49.
34 Quoted in Westergren 1880, 67; see also Neudörfer 1874 in Roussel 1876a, 158–160.
35 Roussel 1874, 167; Roussel 1875b, 320–325; Roussel 1885, 22–24.
36 Chadwick 1874, 26.
37 Bartrip 1990, 9. See also Stöckel et al. 2012.
38 Billroth 1869, 14.

3. Blood on the battlefield

1   Cf. Lefrère & Danic 2009, 1011.
2   Niemeyer 1874, 60. Another version of this image, presented as the first painting of a blood transfusion, can be found on the website of the Finnish Red Cross Blood Service, see: https://www.bloodservice.fi/about-us/history.
3   Roussel 1876a, 152.
4   Roth 2016/1990, 465–466.
5   Wawro 1996, 65–84.
6   Wawro 1996, 123.
7   Becker 1997, 658.
8   Hess 1997, 482.
9   Hobsbawm 1977, 99.
10 Roussel 1876a, 151.
11 Wawro 1996, 150.
12 Hess 1997, 482.
13 Neudörfer 1860, 124–130, 142–147.
14 *Bulletin Général de Thérapeutique médicale et chirurgicale* 1860, 578.
15 Barnes 1883, 811–812.
16 Lidell 1888, 556.
17 E. g. Fischer 1867, 78; Beck 1867, 120–123.
18 de Santi & Dziewonski 1882, 941.
19 According to Roussel 1885, 32; see also Billroth 1869, 45.
20 Fischer 1868, 294–296.
21 McCallum 2008, 126.
22 Showalter 2004, 293.
23 Swain 1970, 514.
24 Showalter 2004, 293; Gabriel 2013, 179–183; Garrison 1970/1922, 178–179.
25 Showalter 2004, 292–293; Gabriel 2013, 181.
26 Hasse 1875, 248.
27 For a vivid description of the situation for typhus patients at the German military hospital at Ecouen near Paris, see Forbes 1871, 73–78. Forbes was special correspondent to *The Daily News*.
28 Hasse 1875, 272.
29 Köhler 1890, 342–347.
30 Jullien 1875, 303–305.
31 de Santi & Dziewonski 1882, 994.

32 Bruberger 1874, 533.
33 Gesellius 1873, 183.
34 Gesellius 1874, 14–15.
35 Gesellius 1874, 15.
36 Gesellius 1874, 15.
37 Oré 1876, 218.
38 Albini 1872, 258.
39 Hasse 1875, 249.
40 Schliep 1874, 27.
41 Heyfelder 1875b, 303.
42 Eckert 1876, 159.
43 Gesellius 1874, 159.
44 Eckert 1876, 166.
45 Gesellius 1874, 11.
46 Eckert 1876, 160.
47 Eckert 1876, 166.
48 Neudörfer 1875b, 61–62, 103–104.
49 Roussel 1876a, 151.
50 See e.g. Icard 1905.
51 Von Nussbaum, quoted in Chenu 1871, vi.
52 Neudörfer 1872, 1408.
53 Eckert 1876, 163–164.
54 Roussel 1876a, 151.
55 Neudörfer 1874 in Roussel 1876a, 158–160.
56 Hasse 1875, 248–249.
57 Fischer 1882, 902.
58 Bruberger 1874, 532–533.
59 [Seyfer]th 1874, 46.
60 – (1876), 365.
61 Schliep 1874, 26–28.
62 Cooter 2004, 333–334.
63 Cooter 2004, 344.
64 Edholm 1876, 9.

4. Blood for the lungs

1 Torstensson 1874a.

2   Torstensson 1874b, 674.
3   Torstensson 1874b, 674–675.
4   Torstensson 1874b, 677.
5   Warfvinge 1876, 152.
6   Westerberg 1875, 139–141.
7   Daniel, Thomas M. 2006, 1864; see also Bynum, Helen 2012.
8   Hasse 1874a, 62–63. Hasse was not the first to try animal blood against phthisis. In 1839, Heinrich Bliedung of Kirchbarkau in Schleswig-Holstein twice injected buck's blood into the veins of a man suffering from phthisis. The operation was deemed successful but had no followers. See – (1841), 320–321.
9   Sander 1874, 174.
10 Wallis 1876, 36.
11 Hotz 1875, 21.
12 Merkel 1875, 91–97; – (1874g); *Marysville Daily Appeal*, February 10, 1875; Sittel 1874, 157–158, 169–170, 183–184.
13 – (1874a); – (1874b); – (1874d); – (1874h); – (1875). Dr Dessauer at the German hospital in Valparaiso, Chile, made three transfusion of lamb's blood to phthisis patients; the results were unfavourable. See Morton 1877, 17.
14 Sander 1874, 173–175, 189–191.
15 Heyfelder 1875a, 111.
16 Flemming 1874, 654–656; Warfvinge 1876, 151; Thurn 1874a, 393–395; Küster 1874, 385–404; Brügelmann 1874, 395–397, 423–425; Neudörfer 1875b, 85–97; Schmidt 1874, 137–141, 147–151; Schmidt 1875, 106–108; Molitor 1875, 75–80; Klingelhöffer 1874, 419–422, 446–449.
17 See e.g. Madge 1874, 42–44. Also: Pelis 2001a, 187–188.
18 Aveling 1874/75, 224.
19 Accounts differ about whether Dr C. Faber's transfusion at the German Hospital was a success or not. See *The Lancet*, September 26, 1874, 462 and October 10, 1874, 541 and *Proceedings of the Medical Society of London* 1874/75, pp. 17.
20 *Jönköpingsposten*, September 2, 1874; *Borås tidning*, July 4, 1874.
21 Fiedler & Birch-Hirschfeld 1874, 546.
22 See e.g. –(1874b); – (1874c); – (1874e). A similar story was told from Hôtel Dieu in Paris in 1876. A certain Jean Scipion fell seriously ill, no human blood donor was to be found, wherefore he was transfused with the blood of a sheep. He recovered, but had delusions, was convinced he had turned into a sheep himself, bellowed, and got extremely scared

whenever he passed a butcher's shop. The case was to be examined by the Academy of Medicine. See L'Echo Saumurois, February 4, 1877.
23 Gesellius 1874, 15–16.
24 One who did try was Carl Hueter, professor of surgery in Greifswald and a critic of lamb blood transfusion. See, for example, Hueter 1876, 161–187.
25 Chadwick 1874, 30.
26 Neudörfer 1875b, 49. For an amusing depiction of fainting blood-donors, see Jennings 1883, 3.
27 Hurwitz 2006; Rylance 2006; Stowe 1996.
28 E.g. Warfvinge 1878, 133–140.
29 Hasse 1874a.
30 Hasse 1874a, 21, 25.
31 Klingelhöffer 1874, 448.
32 Waldenström 1875, 192; Petersson 1875b, 186.
33 Flemming 1874, 654–656.
34 Hasse 1875, 275.
35 Cradle 1874, 295.
36 Neudörfer 1875b, 57–58.
37 Cf. Hasse 1874b, 405–410.
38 Hasse 1874a, 74.
39 Björck 1875, 173.
40 Björck 1875, 174.
41 Torstensson 1874b, 675.
42 Petersson 1875a, 179.
43 Cradle 1874, 295.
44 Björck 1875, 175; Torstensson 1874b, 676.
45 Björck 1875, 175.
46 – (1874i), 155.
47 Gissler & Wentzel 1874, 70; Merkel 1875, 94–95.
48 Heyfelder 1875b, 309.
49 – (1874h).
50 Beigel 1874, 493.
51 Roelen 1874, 35.
52 Björck 1875, 175–176.
53 Svensson 1875, 331.
54 Petersson 1875a, 180–181.
55 Braun, citing Roussel, see Braun 1878, 90.
56 Hasse 1875, 269.

57 E. g. Svensson 1875, 337; Panum 1875a, 85–87.
58 Hasse 1875, 269.
59 Redtel 1873/74, 581.
60 Merkel 1875, 95.
61 Masing 1873/74, 72.
62 Hasse 1875, 267.
63 Hasse 1874a, 43.
64 Hasse 1874a, 51.
65 v. Cube 1877, 378–381.
66 Brügelmann 1874, 425.
67 Redtel 1873/74, 583.
68 Penzoldt 1874, 502.
69 Stern 1874, 56–57.
70 Svensson 1875, 334.
71 Fiedler & Birch-Hirschfeld 1874, 565.
72 Heyfelder 1875a, 113.
73 Thurn 1874b, 659.
74 Fiedler & Birch-Hirschfeld 1874, 574.
75 Hasse 1875, 291.
76 Röhlen 1875, 48; Fiedler & Birch-Hirschfeld 1874, 591.
77 Wallis 1876, 35.

5. Asylum experiments

1 Laguzzi 1874, cited in Ponza 1875, 54.
2 The physiologist Pierre Cyprien Oré in Bordeaux performed a lamb blood transfusion on a pellagrous patient in June 1876, but with fatal results; see *Presse Médicale Belge*, 1876, 295. At about the same time, Dr Voisin at La Salpétrière in Paris had two patients under treatment with injections of lamb's blood; no results were reported. See *Medical Times and Gazette*, January 29, 1876, 127. An unsuccessful attempt to transfuse a schizophrenic patient was made in Germany, see Leube 1874, 95–96.
3 See Tucker 1887, 1207–1309; Tagliavini 1985, 175.
4 Calculated from accounts in contemporary Italian medical journals and Oré 1876.
5 *Göteborgsposten*, February 25, 1880.

6 Tucker 1887, 1270. For contemporary analyses of the history, causes, symptoms and treatments of pellagra, see Roussel, T. 1866; Lombroso, 1870.
7 See Whitaker 1992, 80–90; Gentilcore 2014, 48–54; Priani 2017, 166–181.
8 Ginnaio 2011, 673.
9 Ginnaio 2011, 674. Contemporary accounts of this strange Italian disease were found in newspapers as far away as Sweden; see *Dagens Nyheter*, November 23, 1878 and *Göteborgsposten*, February 25, 1880.
10 Priani 2017. For a discussion of the Italian nomenclature for mental disease in the 19th century, see de Fazio 2015.
11 Livi 1875, 73.
12 *The Journal of Mental Science* 1876, 11–12.
13 Ginnaio 2011, 586.
14 Whitaker 1992, 83.
15 Gentilcore 2014, 53–54; Mariani-Costantini & Mariani-Costantini 2007, 167.
16 E.g. Mariani-Costantini & Mariani-Costantini 2007, 168–169.
17 Ginnaio 2011, 679.
18 Brocca 1875, 390–398.
19 See Leidesdorf's report at the meeting on February 6, 1874 of the *Gesellschaft der Aerzte in Wien*, referred to in the *Allgemeine medicinische Central-Zeitung* 1874, 166–167. Roussel 1877, 78–82 gives a detailed account of the Vienna transfusion made with his apparatus. He also derides the Italians who followed this example but with a different and, in his view, deficient instrument.
20 Summarized from Tamburini 1874, 186–196, and archival material at the San Lazzaro Asylum, Reggio Emilia.
21 *Archivio Italiano ...* 1874, 364.
22 *Archivio Italiano ...* 1874, 360.
23 *Archivio Italiano ...* 1874, 322.
24 Quoted in Jullien 1875, 176.
25 See e.g. de Cristoforis 1875; Morselli 1876; Schivardi 1874, 289–290.
26 See e.g. Vizioli 1874, 673–685.
27 Zimmerman & Howell 1932, 419–429; Marinozzi et al. 2018, 1–5.
28 Tagliavini 1985, 178–183; Reeder 2012, 192–193.
29 Manzini & Rodolfi 1876, 154.
30 Sutherland 1855, 261–288.
31 Schivardi 1874, 290; Ponza 1875, 28; Manzini & Rodolfi 1876, 161.

32 Based on Livi 1875, 82–88, and archival material, San Lazzaro Asylum, Reggio Emilia.
33 Ponza 1875, 26–56. I have here used the more detailed account reported by Ponza's collaborator Pacchiotti in *L'Union Médicale* 1874, 342–343.
34 Michetti 1875, 21–22.
35 *Archivio Italiano...* 1874, 377–378. I have slightly edited the translation published in the *London Medical Record*, Jan. 20, 1875, 44.
36 It was a Cagnola Foundation prize. See: – (1872), 1040. The academy was founded by Napoleon in 1797; its first president was Alessandro Volta.
37 Polli 1852. For Polli's argument for the competition and the formulation of the call, see Polli 1872, 1056–1058.
38 – (1875), 796–799.
39 de Cristoforis 1875a; Lombroso 1876, 665–686, 737–831.
40 Manzini & Rodolfi 1875, 343–370.
41 Manzini & Rodolfi 1876, 111–116, 131–135, 141–144, 151–155, 161–162. Their accounts, as well as those by de Cristoforis and Lombroso, were published after the conclusion of the competition. Manzini and Rodolfi include cases up to August 1875.
42 http://www.enciclopediabresciana.it/enciclopedia/index.php?title=MANZINI_Giovanni_Battista.
43 Cited in Luigi & Vernia 1997, 16–17.
44 Rodolfi 1874, 27–36.
45 Cascella 2015, 1593–1597.
46 Manzini & Rodolfi 1876, 115.
47 The same proportion as in the asylum population, see Manzini 1877, 202. For a history of the Brescia asylum in the 19th century, see Porro 2005, 75–94.
48 Manzini & Rodolfi 1876, 115.
49 Manzini & Rodolfi 1876, 115.
50 Manzini & Rodolfi 1876, 112.
51 Manzini & Rodolfi 1876, 114.
52 Manzini & Rodolfi 1876, 144.
53 Manzini & Rodolfi 1876, 144.
54 Manzini & Rodolfi 1875, 369.
55 Calculated from the accounts in Manzini & Rodolfi 1875 and 1876.
56 Manzini & Rodolfi 1876, 132
57 Manzini & Rodolfi 1875, 346.
58 Manzini & Rodolfi 1875, 348, 354.

59 Raz 2017.
60 Manzini & Rodolfi 1876, 153.
61 Manzini & Rodolfi 1876, 154–155.
62 See e.g. Verga 1875, 415.
63 Quoted by Tagliavini 1985, 183.
64 Manzini & Rodolfi 1876, 154 (edited by me).
65 Manzini & Rodolfi 1876, 154.
66 Zannetti 1876, 78.
67 Bergonzi 1875, 56–60.
68 Bozzolo 1875, 551–557; see also Morselli 1876, 560, for the critique by the physiologist Paolo Mantegazza.
69 Zannetti 1876, 78; Schivardi 1874, 289.
70 Livi 1875; Lombroso 1876, 665–686, 737–831; Morselli 1876.
71 Cited by Tamburini 1874, 195–196.

6. Proofs and refutations

1   Landois 1875, 271–273.
2   Landois 1875, 159.
3   Starr 1998, 38–40.
4   Ponfick 1874a; Ponfick 1874b; Worm-Müller 1875.
5   Panum 1875a; Panum 1875b; Panum 1876.
6   Jullien 1875, 216.
7   Wallis 1876, 36.
8   See Jewson 1976, 225–244.
9   Mayer 1874, 271.
10 Redtel 1873/74, 581. See also Müller 1955, 202–205.
11 Bleker 1997, 23.
12 Mayer 1874, 272.
13 Panum 1864.
14 Panum 1876, 53–54.
15 Panum 1875b, 2.
16 Panum 1875b, 73.
17 Statistisches Jahrbuch 1963, 57.
18 Panum 1875b, 74.
19 Jewson 1976, 228.
20 Bynum, W. 1994, 60–61; Bynum, W. 2008, Chapter 4.

21 Schlich 2004, 70.
22 Schlich 2004, 70.
23 Risse 2015, 9.
24 Volz 1870 as quoted in Risse 1982, 39.
25 Rosenberg 1977, 500.
26 Lesser 1877, 238 (slightly edited by me). Original German version 1874.
27 Wurtz 1870. For Great Britain, see Pelis 1997.
28 – (1874f), 107.
29 Schorr 1956, 38–40.
30 Landois 1875, 279–280; 198–200. See also Sanderson 1873, and the discussion in Richards 1986.
31 Warfvinge 1876, 134.
32 Panum 1876, 51–52.
33 Rossander 1875, 152–153.
34 E.g. Lovén 1876, 88.
35 Hasse 1875, 267.
36 Ponfick 1874b.
37 Rossander 1875, 157–158.
38 von Linhart 1874, 99.
39 Neudörfer 1875b, 112.
40 Roussel 1876a, 326.
41 *The Doctor*, 1875, 144.
42 Oré 1876, 235–241.
43 Rossander 1875, 158.
44 Jahn 1874, 9.
45 Svenska läkaresällskapets förhandlingar 1875, 28.
46 von Cube cited in *Jahresbericht über die Leistungen und Fortschritte Gesammten Medicin* 1877, 330.
47 Svensson 1875, 333–334.
48 von Cube 1877, 378.
49 Fiedler & Birch-Hirschfeld 1874, 565. For the history of comparisons in clinical experiments, see Chalmers 2001.
50 Panum 1875a, 85–87.
51 Lesser 1877, 238.
52 Morselli 1876, 521–523; Bozzolo 1875, 551–557; de Cristoforis 1875b, 557–558.
53 Landois 1875, 291.
54 Lindes 1875, 161.

55 Mayer 1874, 272. See also Björck 1875, 177; Roussel 1875a, 163; Hasse 1875, 291.
56 Martin 1859; Oré 1868; von Belina-Swiontkowsky 1869; Marmonnier 1869; Sacklén 1870; Gesellius 1873.
57 Jahn 1874, 9.
58 Hasse 1874b, 409.
59 Oré 1876, 241.
60 Landois 1875, 327–356; Hirschfeld & Hirschfeld 1976, 471.
61 Wallis 1876, 23–31.
62 Panum 1875b, 2.
63 Lombroso 1876, 774–795.
64 Roussel 1875a, 132–133.

7. Transgressions

1   von Weber 1879.
2   Panum 1879, 25.
3   Birke & Michael 1998, 245–261; Brown 1999, 327–355.
4   Landois 1875, 271–275.
5   Hasse 1875, 285; Panum 1876, 53–54.
6   Gould 1977.
7   Manzini & Rodolfi 1876, 151; Neudörfer 1875b, 67.
8   Hasse 1874a, 64–65; Masing 1873/74, 74; Roelen 1874, 35; Heyfelder 1875a, 115; Schivardi 1874, 298; Eckert 1876, 113.
9   Panum 1864; Landois 1875.
10 Mantegazza 1880. See Bretschneider 1962, 99, 121; Guerrini 2003, 80–81.
11 But see Armstrong 2016, 138 for experiments on sheep within agricultural research.
12 Cunningham & Williams 1992, 8. Cf. Richards 1986.
13 See Logan 2002, 329–363; Rader 2004.
14 Descola 2005.
15 Descola 2005, 168–180; Schlich et al. 2009.
16 Rémy 2009, 407.
17 Quoted in Sahlins 2015, 31–32.
18 Douglas 1966.
19 As referred in Björnström 1873/74, 180.
20 Barnes 1876, 577.

21 Quoted from a debate in the British Medical Association, *The Obstetrical Journal of Great Britain and Ireland* 1873/74, 543.
22 Madge 1874, 43.
23 de Cristoforis 1875a, 91.
24 Eyselein 1874, 301.
25 Steiner 1874, 309.
26 Quoted in Schivardi 1874, 290–291.
27 Pacchiotti 1874, 237–238.
28 Maehle 2009, 78.
29 Svensson 1875, 334.
30 Cf. Wilde 2009, 302–330.
31 Sittel 1874, 158.
32 Westerberg 1875, 140.
33 Halpern 2004, 4.
34 Maehle 2009, 70.
35 See Pellegrino & Thomasma 1987.
36 Neudörfer 1875a, 564–568.
37 Chadwick 1875, 38 (one paragraph in original).
38 Quoted in Fennell 1996, 42–43.
39 Dyck & Stewart 2016, 8.
40 Here quoted from the English translation in Bernard 1949, 102.
41 Bynum, W. 1988, 35; Guerrini 2003.
42 Pacchiotti 1874, 237.
43 Fennell 1996, 39–41.
44 *Le Salute*, Genua, September 19, 1874, cited in *Le Progrès Médical* 1875, 16.
45 Engdahl 1874, 508.
46 Trendelenburg 2011, 129–133.
47 Panum 1875a, 90.
48 Quoted in Vizioli 1875, 176.
49 Morselli 1876, 179.
50 Tschin 1874, 1169–1170.
51 *Wiener Medizinische Wochenschrift* 1874, 1094.
52 *Allgemeine Wiener Medizinische Zeitung* 1874, 444.
53 Bergonzi 1875, 60.
54 Schmidt 1875, 107–108.
55 Landois 1875, iv.
56 That is, from a song of praise to a swan song. Bruberger 1875, 214.

8. Winding up

1. Ljungström 1996, 81–83.
2. *Kalmar*, May 22 1876, 3.
3. Tillhagen 1989, 275; Stapelberg 2016, 92–94.
4. Recorded at Fränninge 1933. Folklore Archive, Lund, Accession number M 3774.
5. Pettersson 1976, 182–199; Zimmerman & Howell 1932, 412–413.
6. *Stockholms Korrespondens*, June 10, 1875.
7. Dallera 1876, 117–123.
8. Warfvinge 1878, 133–140; Girerd 1881, 4–13; – (1879), 7.
9. von Bergmann 1883, 4.
10. von Bergmann 1883, 8–13.
11. von Bergmann 1883, 23.
12. Hayem 1882, 486.
13. A final state-of-the-art overview was given by Haehner 1880.
14. Pelis 2001a, 185–213; Jennings 1883.
15. Panum 1864; Panum 1875a; Hayem 1882.
16. Rossander 1875, 158.
17. Pelis 2001a, 194–195.
18. Pelis 2001b, 242.
19. Jenkins 1890, 446–447.
20. Elkeles 1996, 76–89.
21. Hasse 1875, 292.
22. Müller 1955, 203.
23. Daniel, Diana 2013, 69–70; Grosberg 1926, 10.
24. See the comments in *La presse médicale Belge*, 1876, 327–328; 368–373.
25. Roussel 1882, cover page.
26. Roussel 1876b, 29–31.
27. Roussel 1877, 96.
28. Roussel 1885, 56.
29. de Santi & Dziewonski 1882, 946.
30. *L'Union médicale* 1876, 614.
31. Roussel 1882, 23.
32. Delorme 1888, 527–528.
33. Roussel 1895.
34. Olivier 1952, 9–10.
35. Köhler 1906, 296–306.

36   Köhler 1890, 346.
37   von Esmarch 1894, 116.
38   Hertzberg 1869, 24.
39   Pelis 2001b, 238–277; Schneider 1997, 105–126.

Epilogue: The return

1   Bernheim 1891, 192.
2   Lahaie & Watier 2017, 776–777.
3   – (1891a), 1.
4   – (1891b); – (1891c); Lahaie & Watier 2017, 777.
5   Simon 2007, 63–82.
6   *Centralblatt für innere Medicin* 1894, 372.
7   *Annual of the Universal Medical Sciences* 1894, A-95.
8   Bier 1901, 572.
9   Bier 1921, 163.
10 Blessing 2011, 64–65.
11 v. Klebelsberg 1922, 611–626.
12 Zimmer & v. Balden 1931, 244.
13 Hadenfeldt 1931, 235.
14 Famius 1916, 330.
15 Cruchet 1926, 975.
16 – (1928), 689.
17 San Lazzaro Asylum Archive, Reggio Emilia.
18 Kirchenbücher, Schwenda.
19 Swedish National Archive, Arkiv Digital.

Sources and Literature

Archives

Folklivsarkivet (Folklore Archive), Lund, Sweden.
Kirchenbücher, Schwenda, Evangelisches Kirchspiel am Auerberg, Germany.
Archivio ex Ospedale psichiatrico San Lazzaro, Reggio Emilia, Italy.
Stadtarchiv, Nordhausen, Germany.
Swedish National Archive, Arkiv Digital.
Universitätsarchiv, Greifswald, Germany.

Websites

https://nordhausen-wiki.de/index.php?title=Oscar_Hasse (accessed October 20, 2019).
https://www.google.com/search?client=firefox-b-d&q=sanguine+definition (accessed October 20, 2019).
https://www.nordhausen.de/allgemein/cblock_lang.php?cBlNr=11508 (accessed October 20, 2019).
https://www.bloodservice.fi/about-us/history (accessed October 20, 2019).
http://www.enciclopediabresciana.it/enciclopedia/index.php?title=MANZINI_Giovanni_Battista. (accessed October 20, 2019).

Literature

– (1834): 'Transfusion of Blood in Uterine Hemorrhage.' In: *The Lancet* October 25, pp. 156–157.

- (1841): 'Ein Fall von Transfusion.' In: *Schmidt's Jahrbücher der in- und ausländischen gesammten Medicin* vol. 130, pp. 320–321.
- (1872): 'Adunanza del 21 novembre 1872.' In: *Rendiconti. Reale Istituto Lombardo di scienze e lettere* vol. 5, pp. 1040.
- (1874a): 'Transfusion of Blood Successfully Performed.' In: *Jeffersonville Evening News* July 7.
- (1874b): 'Transfusion of Blood – An Astonishing Result.' In: *Sacramento Daily Union* July 15.
- (1874c): 'Transfusion of Blood – An Astonishing Result.' In: *Milton Bruce Herald* September 8.
- (1874d): 'Transfusion of Blood.' In: *Richmond Palladium* September 23.
- (1874e): 'En amerikansk kur.' In: *Blekingsposten* October 27.
- (1874f): 'The Present State of Transfusion.' In *Medical Times and Gazette* November 14, pp. 107.
- (1874g): 'Transfusion of Blood. An Interesting Surgical Experiment.' In: *Pacific Rural Press* November 21.
- (1874h): 'Transfusion of Blood.' In: *The Staffordshire Daily Sentinel* December 23.
- (1874i): 'Reference to a Report by Leube.' In: *Deutsche Zeitschrift für Praktische Medizin* nr. 17, pp. 155.
- (1875): 'Rapporto della Commissione, Premj ordinarj di Fondazione Cagnola. Sulla trasfusione del sangue.' In: *Rendiconti. Reale Istituto Lombardo di scienze e lettere*, pp. 796–799.
- (1876): 'Review.' In: *Jahresbericht über die Leistungen und Fortschritte in den Gesammten Medicin*, pp. 365.
- (1879): 'Transfusion of Blood.' In: *Ames Intelligencer* July 25.
- (1891a): 'Another Cure for Consumption.' In: *Oamaru Mail* March 3, pp. 1.
- (1891b): 'Le sang de chèvre.' In: *L'Impartial* March 10, pp. 1.
- (1891c): 'One of the Latest Remedies for Tuberculosis.' In: *The Argus* April 21, pp. 7.
- (1928): 'La Transfusion du sang de l'Animal à l'Homme. By René Cruchet, A. Ragot, and J. Caussimon.' In: *The British Journal of Surgery* vol. XV, pp. 689.

Albini, Giuseppe (1872): 'Relazione sulla trasfusione diretta di sangue d'agnello praticata due volte in una Signora.' In: *Rendiconto dell'Accademia delle scienze fisiche e matematiche* vol. XI, pp. 258–273.

Allgemeine Medicinische Central-Zeitung vol. 43 (1874), pp. 166–167, 275.

Allgemeine Wiener Medizinische Zeitung vol. 19 (1874), pp. 444.

Annual of the Universal Medical Sciences vol. 5 (1894), pp. A-95.

Archivio Italiano per le Malattie nervose e particolarmente per le Alienazioni mentali vol. 11 (1874), pp. 322, 360, 364, 377–378.

Archivio Italiano per le Malattie nervose e particolarmente per le Alienazioni mentali vol. 12 (1875), pp. 56/57.

Armstrong, Philip (2016): Sheep, London: Reaktion.

Aveling, J. H. (1874/75): 'Immediate Transfusion of Lamb's Blood.' In: *The Obstetrical Journal of Great Britain and Ireland* vol. 11, pp. 223–224.

Barnes, Joseph K. (ed.): (1883): 'The Medical and Surg. History of the War of the Rebellion.' In: *Surgical History* vol. 2, pp. 811–812

Barnes, Robert (1876): Lectures on Obstetric Operations, London: J. & A. Churchill.

Bartrip, Peter W. J. (1990): Mirror of Medicine: A history of the British Medical Journal, Oxford: Clarendon.

Beck, Bernhard (1867): Kriegs-chirurgische Erfahrungen während des Feldzuges 1866 in Süddeutschland, Freiburg: Fr. Wagner.

Becker, Annette (1997): 'War Memorials: A Legacy of Total War.' In: Stig Förster/Jörg Nagler (eds.), On the Road to Total War. The American Civil War and the German Wars of Unification, 1861–1871, Cambridge: Cambridge University Press, pp. 657–680.

Beigel, H. (1874): 'Eine Bluttransfusion vom Lamm zum Menschen.' In: *Wiener Medizinische Wochenschrift* vol. 24, pp. 492–493.

Bergonzi, Guiseppe (1875): 'Brève appunti intorno alla trasfusione del sangue.' In: *Archivio Italiano per le Malattie nervose e particolarmente per le Alienazioni mentali* vol.12, pp. 56–60.

Bernard, Claude (1949): An Introduction to the Study of Experimental Medicine, New York: Henry Schuman.

Berner, Boel (2012): Blodflöden. Blodgivning och blodtransfusion i det svenska samhället, Lund: Arkiv.

Bernheim, S. (1891): 'Transfusion du sang de chèvre et tuberculose.' In: *Le Moniteur médical* March 21, pp. 187–195.

Bier, August (1901): 'Die Transfusion von Blut, inbesonders von fremdartigen Blut, und ihre Verwendbarkeit zu Heilzwecken von neuen Gesichtspunkten betrachtet.' In: *Münchener Medizinische Wochenschrift* vol. 48, pp. 569–572.

Bier, August (1921): 'Heilentzündung und Heilfieber mit besonderer Berücksichtigung der parentalen Proteinkörpertherapie.' In: *Münchener Medizinische Wochenschrift* vol. 68, 163–168.

Billroth, Theodor (1869): Die allgemeine chirurgische Pathologie und Therapie, Berlin: Georg Reimer.

Billroth, Theodor (1875): 'Zur Diskussion über einige chirurgische Zeit- und Tagefragen.' In: *Wiener Medizinische Wochenschrift* vol. 25, pp. 1–6, 25–29, 41–43, 65–69.

Birke, Lynda/Michael, Mike (1998): 'The Heart of the Matter: Animal Bodies, Ethics, and Species Boundaries.' In: *Society and Animals* 6/3, pp. 245–261

Björck, C.H. (1875): 'Kronisk pneumoni behandlad med transfusion af D:r O. Torstensson.' In: *Upsala Läkareförenings Förhandlingar* vol. 10, pp. 172–177.

Björnström, Fredrik (1873/74): 'Geselii method för transfusion.' In: *Upsala Läkareförenings Förhandlingar* vol. 9, pp. 180–181.

Bleker, Johanna (1997): 'To Benefit the Poor and Advance Medical Science.' In: Manfred Berg /Geoffey Cocks (eds.), Medicine and Modernity. Public health and medical care in nineteenth- and twentieth century Germany, Washington & Cambridge: Cambridge University Press, pp. 17–34

Blessing, Bettina (2011): Pathways of Homoeopathic Medicine, Springer Media.

Blundell, James (1828/29): 'Observation on Transfusion of Blood.' In: *The Lancet* June 13, pp. 321–324.

Borås tidning, July 4, 1874.

Bozzolo, Camillo (1875): 'Sugli esperimenti di trasfusione sanguigna eseguiti nell'Ospedale Maggiore di Milano dai signori dottori Ponza e Rodolfi.' In: *Annali Universali di Medicina e Chirurgia* vol. 232, pp. 551–557.

Braun, E. (1878): 'J. Roussel: La Transfusion, première serie, trentecinq operations.' In: *Tidskrift i militär helsovård* vol. 1, pp. 89–94.

Bretschneider, H. (1962): Der Streit um die Vivisektion im 19. Jahrhundert, Stuttgart: Gustav Fischer.

Brocca, Giovanni (1875): 'Sulle pellagrose curate nel comparto delle deliranti dell'Ospitale Maggiore di Milano, l'anno 1874.' In: *Archivio Italiano per le Malattie nervose e particolarmente per le Alienazioni mentali*, vol. 12, pp. 390–398.

Brown, Nik (1999): 'Xenotransplantation: Normalizing Disgust.' In: *Science as Culture* 8/3, pp. 327–355.

Bruberger, Dr (1874): 'Über Transfusion und ihren Werth im Felde.' In: *Deutsche Militärärztliche Zeitschrift* vol. 3, pp. 525–534.

Bruberger, Dr (1875): 'Zur Transfusionsfrage.' In: *Deutsche Militärärztliche Zeitschrift* vol. 4, pp. 210–216.

Brügelmann, W. (1874): 'Ein Fall von Phthisis pulmonum, durch Inhalationen und eine Lammbluttransfusion geheilt.' In: *Berliner klinische Wochenschrift*, vol.11, pp. 395–397, 423–425.

Buess, H. (1953): 'Der Ausbau der Bluttransfusion in neuester Zeit.' In: *Bulletin der schweizerischen Akademie der Wissenschaften* vol. 9, pp. 248–269.

Bulletin Général de Thérapeutique médicale et chirurgicale vol. 59 (1860), pp. 578.

Bynum, Helen (2012): Spitting Blood. The History of Tuberculosis, Oxford: Oxford University Press.

Bynum, W.E. (1994): Science and the Practice of Medicine in the Nineteenth Century, Cambridge: Cambridge University Press.

Bynum, William (1988): 'Reflections on the History of Human Experimentation.' In: Stuart F. Spicker et al. (eds.), The Use of Human Beings in Research, Dordrecht: Kluwer, pp. 29–46

Bynum, William (2008): The History of Medicine. A Very Short Introduction, Oxford: Oxford University Press.

Cascella, Marco (2015): 'The Controversial Experiments on the Intravenous Administration of Drugs (and Air!) during the Cholera Epidemic of 1867 in Italy.' In: *Revista medica de Chile* vol. 143, pp. 1593–1597.

Casse, Joseph (1874): De la transfusion du sang, Bruxelles: Henri Manceaux.

Centralblatt für innere Medicin, 1894, pp. 372.

Chadwick, James R. (1874): 'Transfusion.' In: *The Boston Medical and Surgical Journal* vol. XCI, pp. 25–32.

Chadwick, James R. (1875): 'A Case of Immediate Transfusion.' In: *The Boston Medical and Surgical Journal* vol. XCII, pp. 33–38.

Chalmers, Iain (2001): 'Comparing like with like: Some Historical Milestones in the Evolution of Methods to Create Unbiased Comparison Groups in Therapeutic Experiments.' In: *International Journal of Epistemology* 30/5, pp. 1156–1164.

Chenu, Jean-Charles (1874): Aperçu historique, statistique et clinique sur le service des ambulances et des hôpitaux, Paris: J. Dumaine.

Cooter, Roger (2004): 'Medicine in War.' In: Deborah Brunton (ed.), Medicine Transformed: Health, Disease and Society in Europe, 1800–1930, Manchester: Open University, pp. 331–363.

Cradle, H. (1874): 'Two Cases of Direct Transfusion from Animal to Man.' In: *The Medical Examiner* vol. XV, pp. 294–295.

Cruchet, René (1926): 'Transfusion of Blood from Animal to Man.' In: *The British Medical Journal* November 27, pp. 975–980.

Cunningham, Andrew/Williams, Perry (1992): 'Introduction.' In: Andrew Cunningham/Perry Williams (eds.), The Laboratory Revolution in Medicine, Cambridge: Cambridge University Press, pp. 1–13.
Dagens Nyheter, November 23, 1878.
Daily Alta California, July 2, 1871.
Dallera, Ernesto (1876): 'Osservazioni cliniche intorno ad un caso di trasfusione diretta ed alcune parole in difesa della trasfusione eterogenea.' In: Il Morgagni vol. XVIII, pp. 117–123.
Daniel, Diana (2013): Der Capillarblut-Tranfusor. Eine Nichtwissensgeschichte der Bluttransfusion im 19. Jahrhundert, Magisterarbeit, Berlin.
Daniel, Thomas M. (2006): 'The History of Tuberculosis.' In: Respiratory Medicine 100/11, pp. 1862–1870.
de Cristoforis, Malachia (1875a): La trasfusione del sangue, Milano: Fratelli Rechiedei.
de Cristoforis, Malachia (1875b): 'Commentatore.' In: Annali Universali di Medicina e Chirurgia vol. 232, pp. 557–558.
de Fazio, Debora (2015): 'I nomi della follia. Premesse per un'indagine storica sulla terminologia della psichiatria in Italia.' In: Zeitschrift für romanische Philologie 131/2, pp. 483–510.
de Santi/Dziewonski (1882): 'De la transfusion du sang en chirurgie d'armée.' In: Revue de Chirurgie vol.2, pp. 938–954, 1030–1043.
Delorme, E. (1888): Traité de chirurgie de guerre, Paris: Félix Alcan.
Descola, Philippe (2005): Par-delà nature et culture, Paris: Gallimard.
Dieffenbach, J. F. (1828): Die Transfusion des Blutes und die Infusion der Arzeneien in die Blutgefäße, Berlin: Enslin.
Douglas, Mary (1966): Purity and Danger: An analysis of Concepts of Pollution and Taboo, London: Routledge.
Dyck, Erika/Stewart, Larry (2016): 'Introduction.' In: Erika Dyck/Larry Stewart (eds.), The Uses of Humans in Experiment. Perspectives from the 17th to the 20th Century, Leiden/Boston: Brill/Rodopi, pp. 1–27.
Eckert, Joseph Friedrich (1876): Objective Studie über die Transfusion des Blutes und deren Verwerthbarkeit auf dem Schlachtfelde, Wien: Moritz Perles.
L'Echo Saumurois, February 4, 1877.
Edholm, Edw. (1876): 'Några drag ur militärmedicinens utvecklingshistoria, med särskildt afseende vid den militära helsovården framsteg i våra dagar till följd av de stora krigen.' In: Tidskrift i militär hälsovård vol.1, pp. 1–21.

Elkeles, Barbara (1996): Der moralische Diskurs über das Medizinische Menschenexperiment im 19. Jahrhundert, Stuttgart: Gustav Fischer.

Engdahl, Edward (1874): 'Om blodtransfusion.' In: *Hygiea* vol. 36, pp. 506–509.

Eyselein, Dr (1874): 'Dr med. Hasse's 29. Lammbluttransfusion.' In: *Deutsche Zeitschrift für praktische Medicin* nr. 35, pp. 299–301, 307–309.

Famius, Dr (1916): 'La transfusion du sang est une opération anodine.' In: *La Science et la Vie* July 27, pp. 323–331.

Fennell, Phil (1996): Treatment without Consent: Law, Psychiatry and the Treatment of Mentally Disordered People since 1845, London: Routledge.

Fiedler, A./ Birch-Hirschfeld (1874): 'Zur Lammblut-Transfusion.' In: *Deutsches Archiv für klinische Medicin* vol. 13, pp. 545–592.

Fischer, Hermann (1868): Lehrbuch der allgemeinen Kriegs-Chirurgie, Erlangen: Ferdinand Enke.

Fischer, Hermann (1882): Handbuch der Kriegschirurgie 2, Stuttgart: Ferdinand Enke.

Fischer, K. (1867): Militairärztliche Skizzen aus Süddeutschland und Böhmen, Aarau: Sauerländer.

Flemming, Dr (1874): 'Eine directe Lammbluttransfusion.' In: *Allgemeine Medicinische Central-Zeitung* vol. 43, 653–656 ,665–667.

Forbes, Archibald (1871): My Experiences of the War between France and Germany, vol II, Leipzig: Bernhard Tauchnitz.

Gabriel, Richard A. (2013): Between Flesh and Steel: A History of Military Medicine from the Middle Ages to the War in Afghanistan, Washington: Potomac.

Garrison, Fielding H. (1970/1922): Notes on the History of Military Medicine, Washington: Association of Military Surgeons.

Gentilcore, David (2014): 'Peasants and Pellagra in 19th Century Italy.' In: *History Today* September 2014, pp. 48–54.

Gesellius, Franz (1873): Die Transfusion des Blutes. Eine historische, kritische und physiologische Studie, St. Petersburg: Eduard Hoppe.

Gesellius, Franz (1874): Zur Tierblut-Transfusion beim Menschen, St. Petersburg: Eduard Hoppe.

Gilder, S.S.B. (1954): Francesco Folli and Blood Transfusion.' In: *Canadian Medical Association Journal* 71/2, pp. 172.

Ginnaio, Monica (2011): 'La pellagre en Italie à la fin du XIXe siècle: les effets d'une maladie de carence.' In: *Population* 66/3, pp. 671–698.

Girerd, L. (1881): Souvenirs cliniques de chirurgie, Paris: Bureau du Siècle Médical.

Gissler Dr/Wentzel, Dr (1874): 'Eine Lammblut-Transfusion.' In: *Aerztliche Mittheilungen aus Baden* vol. XXVIII, pp. 69–71.

Gould, Stephen Jay (1977): Ontogeny and Phylogeny, Cambridge, Mass: Belknap.

Greenwalt, T. J. (1997): 'A Short History of Transfusion Medicine.' In: *Transfusion* 35/5, pp. 550–563.

Grosberg, Oskar (1926): 'Aus meinen Journalistenleben.' In: *Herdflammen*, February, pp. 9–10.

Guerrini, Anita (2003): Experimenting with Humans and Animals: From Galen to Animal Rights, Baltimore/London: Johns Hopkins University Press.

Hadenfeldt, Hans (1931): 'Über intravenöse Tierblutgaben in der Chirurgie.' In: *Deutsche Zeitschrift für Chirurgie* vol. 234, pp. 228–235.

Haehner, H. (1880): 'Die neuern Mittheilungen über Transfusion des Blutes.' In: *Schimidt's Jahrbücher der in- und ausländischen gesammten Medicin* vol. 187, pp. 81–106, 177–198.

Halpern, Sydney A. (2004): Lesser Harms: The Morality of Risk in Medical Research, Chicago: University of Chicago Press.

Harpers Weekly, Supplement, June 4, 1874.

Hasse, Oscar (1868): 'Aus der chirurgischen Abteilung des Herrn. Geh. R. Wilms im Diakonissenhause Bethanien zu Berlin.' In: *Berliner klinische Wochenschrift* vol. 5, pp. 1–5, 13–17, 25–27, 33–36, 45–47.

Hasse, Oscar (1869): 'Einige Fälle von Transfusion.' In: *Berliner Klinische Wochenschrift* vol. 6, pp. 370–373.

Hasse, Oscar (1874a): Die Lammblut-Transfusion beim Menschen. Erste Reihe: 31 eigene Transfusionen umfassend, St. Petersburg: Eduard Hoppe.

Hasse, Oscar (1874b): 'Ueber das Operations-Verfahren bei der directen Thierblut-Transfusion.' In: *Archiv für Chirurgie*, vol. XVII, pp. 405–410.

Hasse, Oscar (1875): 'Ueber Transfusion. Eine Erwiderung auf Professor P.L. Panum's Abhandlung: Zur Orientirung in der Transfusionsfrage.' In: *Archiv für pathologische Anatomie und Physiologie und für klinische Medicin*, vol. LXIV, pp. 243–292.

Hayem, Georges (1882): Leçons sur les modifications du sang sous l'influence des pratiques thérapeutiques, Paris: Masson.

Hertzberg, Friedrich W. (1869): Die Transfusion des Blutes, Greifswald: F.W. Kunike.

Hess, Earl J. (1997): 'Tactics, Trenches, and Men in the Civil War.' In: Stig Förster/Jörg Nagler (eds.), On the Road to Total War: The American Civil

War and the German Wars of Unification, 1861-1871, Cambridge: Cambridge University Press, pp. 481-496.

Heyfelder, Oscar (1874): 'Rapport médical au ministère de la guerre de S.M. l'Empereur de Russie, présenté par les chirurgiens experts désignés, 2/14 mars, 1874.' In: J. Roussel (1876a): La Transfusion, Extraits des Archives générales de médicine, Paris: Asselin, pp. 160-163

Heyfelder, Oscar (1875a): 'Zur Lehre von der Transfusion (fortsetzung).' In: *Deutsche Zeitschrift für Chirurgie* vol. 5, pp. 108-119.

Heyfelder, Oscar (1875b): Manuel de chirurgie de guerre, Paris: Berger-Levrault.

Hirschfeld, Jan/Hirschfeld, Anne-Marie (1976): 'Tidiga blodtransfusioner.' In: *Laboratoriet* 10, pp. 462-474.

Hobsbawm, Eric (1977): The Age of Capital, 1848-1875, London: Little Brown.

Hotz, F.C. (1875): 'Transfusion of lamb's blood in the human subject.' *The Boston Medical Examiner* January 1, pp. 20-25.

Hueter, Carl (1876): *Kritische-antikritische Wanderungen*, Leipzig: F.C.W. Vogel.

Hurwitz, B. (2006): 'Form and Representation in Clinical Case Reports.' In: *Literature and Medicine* 25/2, pp. 216-240.

Icard, S. (1905): Le danger de la mort apparente sur les champs de bataille, Paris: A. Maloine.

Jahn, Dr (1874): 'Ueber Transfusion.' In: *Deutsche Zeitschrift für Praktische Medicin* nrs. 2, 3, 4, pp. 9-11, 17-20, 25-27.

Jahresbericht über die Leistungen und Fortschritte Gesammten Medicin vol. 11 (1877): pp. 330.

Jenkins, J. H. (1890): 'Transfusion of lamb's blood in typhoid fever.' In: *North Carolina Medical Journal* vol. XXVI, pp. 446-447.

Jennings, Charles E. (1883): Transfusion: Its History, Indications, and Modes of application, New York: Baillière, Tindall, and Co.

Jennings, Charles E. (1896): 'The treatment of Excessive Haemorrhage.' In: *The British Medical Journal* February 8, pp. 331-332.

Jewson, N.D. (1996): 'The Disappearance of the Sick-Man from Medical Cosmology, 1770-1870.' In: *Sociology* 10/2, pp. 225-244.

Jullien, Louis (1875): De la transfusion de sang, Paris: J-B. Baillière et Fils.

Jürgensen, Th. (1885): 'Antiphlogistic Method of Treatment.' In: *von Ziemssen's Handbook of General Therapeutics*, vol. II, pp. 165-366.

Jönköpingsposten, September 2, 1874.

Kalmar, May 22, 1876.

Klingelhöffer, Dr (1874): 'Vier Fälle von Transfusion am Menschen.' In: *Berliner klinische Wochenschrift* vol. 11, pp. 419–422, 446–449.

Köhler, A. (1890): 'Blutüberleitungen (Transfusionen).' In: *Sanitäts-Bericht über die Deutschen Heere im Kriege gegen Frankreich 1870/71*, Berlin: Mittler und Sohn, pp. 342–347.

Köhler, A. (1906): 'Transfusion und Infusion seit 1830. Mit besondere Berücksichtigung ihrer Verwendung im Kriege.' In: Otto von Schjerning (ed.), Gedenkschrift für v. Leuthold, Berlin: Hirschwald, pp. 296–306.

Küster, Ernst (1874): 'Ueber die directe arterielle Thierblut-Transfusion.' In: *Archiv für klinische Chirurgie*, vol. 16, pp. 385–404.

Laguzzi, Lorenzo (1874): 'L'Agnello della trasfusione.' *Avvisatore*, nr. 86, 1874 (cited in G.L. Ponza (1875): La trasfusione del sangue negli alienati, *Archivio Italiano per le Malattie nervose e particolarmente per le Alienazioni mentali* vol. 12, pp. 54.

Lahaie, Yves-Marie and Hervé Watier (2017): 'Contribution of Physiologists to the Identification of the Humoral Component of Immunity in the 19th Century.' In: *MABS* 9/5, pp. 774–780.

Landois, Leonard (1875): *Die Transfusion des Blutes*, Leipzig: F.C.W. Vogel.

Lefrère, Jean-Jacques and Bruno Danic (2009): 'Pictorial Representation of Transfusion over the Years.' In: *Transfusion* vol. 49, pp. 1007–1017.

Leisrink, Heinrich (1872): Die Transfusion des Blutes, In: Sammlung klinische Vorträge 41. Leipzig: Verlag von Breitkopf & Härtel, pp. 235-246.

Leisrink, Heinrich (1873): 'Die neuen Mittheilungen über Transfusion des Blutes.' In: *Schmidt's Jahrbücher der in- und ausländischen gesammten Medicin*, vol. 157, pp. 265–283.

Le Journal Illustré, May 22, 1892.

Le Progrès Médical, 1875, pp.16.

Lesser, L. (1877): 'Transfusion and autotransfusion.' In: Clinical Lectures on Subjects Concerned with Medicine, Surgery, and Obstetrics by Various German Authors, London: The New Sydenham Society, pp. 238–260.

Leube, W. (1874): 'Fall von Melancholia attonita. Lammbluttransfusion mit guter, aber vorübergehender Einwirkung auf die Geisteskrankheit.' In: *Klinische Beilage zu den Correspondenzblättern des allgem. ärztlichen Vereins von Thüringen* 6, pp. 95–96.

Lidell, John A. (1888): 'Injuries of Bloodvessels.' In: J. Ashhurst Jr. (ed.), The International Encyclopedia of Surgery, vol. II, New York: William Wood, pp. 495–775

Lindes, Dr (1875): 'Referat über einige Lammbluttransfusionen.' In: *St. Peterburger Medicinische Zeitschrift* vol. V, pp. 158–161.

Livi, Carlo (1875): 'La lipemania stupida e la trasfusione del sangue.' In: *Archivio Italiano per le Malattie nervose e particolarmente per le Alienazioni mentali* vol. 12, pp. 60–88.

Ljungström, Jan G. (1996): Skarprättare, bödel och mästerman, Stockholm: Carlsson.

Logan, Cheryl A. (2002): 'Before There Were Standards: The Role of Test Animals in the Production of Empirical Generality in Physiology.' In: *Journal of the History of Biology* 35/2, pp. 329–363.

Lombroso, Cesare (1870): Studj clinici ed esperimentali sulla natura, causa e terapia della pellagra, Milano: Guiseepe Bernardoni.

Lombroso, Cesare (1876): 'Sulla trasfusione del sangue.' In: *Il Morgagni* vol. XVIII, pp. 665–686, 737–831.

London Medical Record, January 20, 1875.

Lovén, Christian (1876): Om blodet. Dess kretslopp och dess betydelse för kroppsväfvnadernas näring, Stockholm: Klemmings.

Luigi, Gian & Vernia, Luciano (1997): Le X giornate di Rodolofo Rodolfi, medico e patriota, *Rocadelle* nr. 17, pp. 16–17.

L'Union Médicale (1874), pp. 342–343.

L'Union Médicale (1876), p. 614.

Madge, Henry M. (1874): 'On transfusion of blood.' In: *The British Medical Journal* January 10, pp. 42–44.

Maehle, Andreas-Holger (2009): Doctors, Honour and the Law: Medical Ethics in Imperial Germany, Basingstoke: Palgrave/McMillan.

Maluf, N.S.R. (1954): 'History of Blood Transfusion.' In: *Journal of the History of Medicine* vol. I, pp. 59–107

Mantegazza, Paolo (1880): Fisiologia del dolore, Firenze: Felice Poggi.

Manzini, G.B. (1877): 'Rendiconti medico-statistico dal 1871 al 1873 sul Manicomio provinicale di Brescia.' In: *Archivio Italiano per le Malattie nervose e particolarmente per le Alienazioni mentali* vol. 14, pp. 201–204.

Manzini, Giov. Bat./Rodolfi, Rodolfo (1875): 'Breve riassunto degli esperimenti eseguiti colla trasfusione del sangue in varii pazzi dei manicomii provinciali di Brescia dall'agosto 1874.' In : *Archivio Italiano per le Malattie nervose e particolarmente per le Alienazioni mentali* vol. 12, pp. 343–369.

Manzini, Giov. Bat./Rodolfi, Rodolfo (1876): 'Sulla trasfusione del sangue.' In: *Gazzetta Medica Italiana-Lombardia* vol. XXXVI, pp. 111–116, 131–135, 141–144, 151–155, 161–162.

Mariani-Costantini, Renato/Mariani-Costantini, Aldo (2007): 'An outline of the history of pellagra in Italy.' In: *Journal of Anthropological Sciences* vol. 85, pp. 163–171.
Marinozzi, Silvia/ Gazzaniga, Valentina/ Iorio, Silvia (2018): 'The Earliest Blood Transfusions in 17th-Century in Italy (1667–1668).' In: *Transfusion Medicine Reviews* 32/1, pp. 1–5.
Marmonnier, Charles (1869): De la transfusion du sang, Paris: Masson.
Martin, Eduard (1859): Über die Transfusion bei Blutungen Neuentbundener, Berlin: Hirschwald.
Marysville Daily Appeal, February 10, 1875.
Masing, E. (1873/74): 'Zwei Transfusionen.' In: *St. Petersburger Medicinische Zeitschrift* vol. IV, pp. 68–78.
Mayer, Dr (1874): 'Zur Transfusions-Frage.' In: *Deutsche Zeitschrift für praktische Medicin* nr. 32, pp. 271–275.
McCallum, Jack Edward (2008): Military Medicine from Ancient Times to the 21st Century, Abc-Clio.
Medical Times and Gazette, January 29, 1876.
Merkel, G. H. (1875): 'Successful transfusion of blood.' In: *The Medical Eclectic* vol. II, pp. 91–97.
Michetti. Antonio (1875): 'Trasfusione del sangue.' In: *Archivio Italiano per le Malattie nervose e particolarmente per le Alienazioni mentali* vol. 12, pp. 18–26.
Molitor, Franz (1875): 'Zwei Lammblut-Transfusionen.' In: *Aerztliche Mittheilungen aus Baden* vol. XXIX, pp. 75–80.
Moore, Pete (2003): Blood and Justice: The 17th Century Parisian Doctor who Made Blood Transfusion History, Chichester: Wiley.
Morselli, Enrico (1876): La trasfusione del sangue, Roma: Ermando Loescher.
Morton, Thomas G. (1877): Transfusion of Blood and its Practical Application, New York: Putnam.
Müller, R. H. Walther (1955): 'Dr med. Oscar Hasse. Leben und Wirken eines Nordhäuser Arztes.' In: *Der Nordhäuser Roland*, pp. 202–205.
Neudörfer, Josef (1860): 'Ueber Transfusionen bei Anaemischen nach langdaurnder Eiterung.' In: *Oesterreichische Zeitschrift für practische Heilkunde* vol. VI, pp. 124–130, 142–147.
Neudörfer, Josef (1872): Handbuch der Kriegschirurgie und der Operationslehre, Zweite Hefte, Leipzig: F.C.W. Vogel.
Neudörfer, Josef (1874): 'Rapport médical au ministère de la guerre de S.M.I.R. l'Empereur d'Autriche-Hongrie, 19 janvier, 1874.' In: J. Roussel (1876a), La

Transfusion, Extraits des Archives générales de médicine, Paris: Asselin, pp. 158–160.

Neudörfer, Josef (1875a): 'Beiträge zur Bluttransfusion I.' In: *Deutsche Zeitschrift für Chirurgie* vol. 5, pp. 537–602.

Neudörfer, Josef (1875b): 'Beiträge zur Bluttransfusion II.' In: *Deutsche Zeitschrift für Chirurgie* vol. 6, pp. 47–112.

Niemeyer, Paul (1874): 'Lebensrettung durch Blutüberleitung (Transfusion).' In: *Daheim* vol. 10, pp. 61–63.

Ny Illustrerad Tidning, December 27, 1879.

Olivier, Jean, (1952): 'Un Genevois précurseur de la transfusion, le Dr Roussel.' In: *La Croix-rouge Suisse* January 11, pp. 9–12.

Oré, Jean-Cyprien (1868): Études historiques et physiologiques de la transfusion de sang, Paris: J-B Baillière et Fils.

Oré, Jean-Cyprien (1876): Études historiques, physiologiques et cliniques sur la transfusion du sang, Paris: J-B Baillière et Fils.

Pacchiotti, G. (1874) : 'La trasfusione del sangue.' In: *Il Baretti* vol. 6. pp. 236–238.

Panum, P.L. (1864): Experimentelle Untersuchungen zur Physiologie und Pathologie der Embolie, Transfusion und Blutmenge, Berlin: Georg Reimer.

Panum, P.L. (1875a): 'Zur Orientirung in der Transfusionsfrage.' In: *Archiv für pathologische Anatomie und Physiologie und für klinische Medicin* vol. LXIII, pp. 1–91.

Panum, P.L. (1875b): 'Til Orientering i Tranfusionsspörgsmålet.' In: *Nordiskt Medicinskt Arkiv*, vol. 7, pp. 1–82.

Panum, P.L. (1876): 'Weitere Bemerkungen zur Orientierung in der Transfusionsfrage.' In: *Archiv für pathologische Anatomie und Physiologie und für klinische Medicin* vol. LXVI, pp. 26–55.

Panum, P.L. (1879): Til Opklaring af nogle Misforstaaelser angaaende Vivisektionen og Dyrebeskyttelsen i Danmark. København: Hagerups.

Pelis, Kim (1997): 'Blood Clots: The Nineteenth-Century Debate over the Substance and Means of Transfusion in Britain.' In: *Annals of Science* 54/4, pp. 331–360.

Pelis, Kim (1999): 'Transfusion, with Teeth.' In: Robert Bud/Bernard Finn/Helmuth Trischler (eds.), Manifesting Medicine: Bodies and Machines, New York: Routledge, pp. 1–30.

Pelis, Kim (2001a): 'Blood Standards and Failed Fluids: Clinic, Lab, and Transfusion Solutions in London, 1868–1916.' In: *History of Science* 39/2, pp. 185–213.

Pelis, Kim (2001b): 'Taking Credit: The Canadian Army Medical Corps and the British Conversion to Blood Transfusion in WW1.' In: *Journal of the History of Medicine* 56/3, pp. 238–277.

Pellegrino, Edmund D./David C. Thomasma (1987): 'The Conflict between Autonomy and Beneficence in Medical Ethics: Proposal for a Resolution.' In: *Journal of Contemporary Health Law & Policy* 3/1, pp. 23–46.

Penzoldt, F. (1874): 'Hasse, Die Lammblut-Transfusion beim Menschen.' In: *Jenaer Literaturzeitung* vol. 1, pp. 501–502.

Petersson, O. V. (1875a): 'Fall af lammblodstransfusion.' In: *Upsala Läkareförenings Förhandlingar* vol. 10, pp. 177–183.

Petersson, O. V. (1875b): 'Fall af hastig död, hvarvid transfusion blef försökt.' In: *Upsala Läkareförenings Förhandlingar* vol. 10, pp. 183–191.

Pettersson, Olof (1976): 'Avrättningar och sjukdomsbot. Några anteckningar till folkmedicinens tankemönster.' In: *Sydsvenska medicinhistoriska sällskapets årsskrift* pp.182–199.

Polli, Giovanni (1852): Ricerche ed experienze sulla trasfusione del sangue, Milano.

Polli, Giovanni (1872): 'Sulla trasfusione del sangue – Motivazione del tema proposto pel concorso al premio Cagnola da conferirsi nel 1875.' In: *Rendiconti. Reale Istituto Lombardo di scienze e lettere* vol. V. pp. 1056–1058.

Ponfick, Emil (1874a): 'Ueber die Wandlungen des Lammblutes innerhalb des menschlichen Organismus.' In: *Berliner Klinische Wochenschrift* vol. 11, pp. 333–336.

Ponfick, Emil (1874b): 'Experimentelle Beiträge zur Lehre von der Transfusion.' In: *Archiv für pathologische Anatomie und Physiologie und für klinische Medicin* vol. V, pp. 273–335.

Ponza, G. L. (1875): 'La trasfusione del sangue negli alienate.' In: *Archivio Italiano per le Malattie nervose e particolarmente per le Alienazioni mentali* vol. 12, pp. 26–56.

Porro, Alessandro (2005): 'Psychiatri e ospedale psichiatrico nel XIX secolo a Brescia.' In: Guiseppe Armocida/Giorgio Bellotti (eds.), Contributi di storia della Psichiatria, Varese: Insubria University Press, pp. 75–94.

Presse médicale Belge (1876), pp. 295, 327–328, 368–373.

Priani, Egidio (2017): '"Shrouded in a Dark Fog": Comparison of the Diagnosis of Pellagra in Venice and General Paralysis of the Insane in the United Kingdom, 1840–1900.' In: *History of Psychiatry* 28/2, pp. 166–181.

Proceedings of the Medical Society of London (1874/75), pp.17.

Rader, Karen (2004): Making Mice: Standardizing Animals for American Biomedical Research, 1900-1955, Princeton: Princeton University Press.
Raz, Carmel (2017): 'Music, Theater, and the Moral Treatment: The Casa dei Matti in Aversa and Palermo.' In: *Laboratoire italien* vol. 20 (http://journals.openedition.org/laboratoireitalien/1581) (Accessed August 1, 2018).
Redtel, Dr (1875/74): 'Case of Transfusion of Lamb's Blood in Pulmonary Consumption.' In: *The Obstetrical Journal of Great Britain and Ireland* vol.11, pp. 580–584.
Reeder, Linda (2012): 'Unattached and Unhinged: The Spinster and the Psychiatrist in Liberal Italy, 1860–1922.' In: *Gender & History* 24/1, pp. 187–204.
Rémy, Catherine (2009): 'The Animal Issue in Xenotransplantation: Controversies in France and the United States.' In: *History and Philosophy of the Life Sciences* 31/3-4, pp. 405–428.
Richards, Stewart (1986): 'Drawing the life-blood of physiology: Vivisection and the Physiologists' Dilemma, 1870–1900.' In: *Annals of Science* vol.43, pp. 27–56.
Riesman, David (1937): 'Bourdelot, a Physician of Queen Christina of Sweden.' In: *Annals of Medical History* 9, pp. 191.
Risse, Guenter B. (1982): 'Patients and Their Healers. Historical Studies in Health Care.' In: N.K. Bell (ed.), Who Decides: Conflicts of Rights in Health Care, Clifton: Springer, pp. 27–45
Risse, Guenter B. (2015): 'The Road to 20th Century Therapeutics: Shifting Perspectives and Approaches.' In: *GBR/ Rd Therapeutics*, pp. 1–23.
Rodolfi, Rodolfo (1874): 'Esperimenti d'injezione di farmachi nelle vene degli uomini e de' bruti.' In: *Commentari dell'Ateneo di Brescia*, pp. 27–36
Roelen, Dr (1874): 'Mittheilung.' In: *Correspondenzblatt der ärzlichen Vereine in Rheinland, Westfalen und Lothringen*, September, pp. 31–42.
Röhlen, Dr (1875): 'Vortrag.' In: *Correspondenzblatt der ärzlichen Vereine in Rheinland, Westfalen und Lothringen*, January, pp. 35–48.
Rosenberg, Charles E. (1977): 'The Therapeutic Revolution: Medicine, Meaning and Social Change in Nineteenth-Century America.' In: *Perspectives in Biology and Medicine* 20/4, pp. 485–506.
Rossander, Carl (1875): 'Experimentela bidrag till läran om transfusion.' In: *Hygiea* vol. 37, pp. 151–159.
Roth, François (2015/1990): La guerre de 1870, Paris: Pluriel.
Roussel, J. (1874): 'Bemerkungen zu dem Aufsatze des Herrn Dr Schliep über directe Thierblut-Transfusion.' In: *Berliner klinische Wochenschrift* vol.11, pp. 167–168.

Roussel, J. (1875a): 'La Transfusion.' In: *Archives générales de médicine* vol. 26/Février, pp. 129–163, 443–462, 690–703.
Roussel, J. (1875b): 'La Transfusion.' In: *Archives générales de médicine* vol. 26/Juillet, pp. 204–210, 320–331, 570–581, 670–687.
Roussel, J. (1876a): La transfusion. Extraits des Archives générales de médicine, Paris: Asselin.
Roussel, J. (1876b): Rapports sur la méthode de transfusion directe, Bruxelles: Henri Manceaux.
Roussel, J. (1877): Transfusion of human blood by the method of J. Roussel, London: J.A. Churchill.
Roussel, J. (1882): Transfusion directe du sang vivant, Paris: Asselin.
Roussel, J. (1884): 'De la transfusion du sang vivant.' In: *Le progrès médical* vol. 12, pp. 810–813, 828–831.
Roussel, J. (1885): Leçons sur la transfusion directe du sang, Paris: Asselin et Houzeau.
Roussel, J. (1895): Médicine hypodermique. Notes pratiques sur l'injection sous-coutanée, Paris: Charaire et Cie.
Roussel, Théophile (1866): Traité de la pellagre et des pseudo-pellagres, Paris: Bureau de l'encyclopédie médicale.
Rylance, R. (2006): 'The Theatre and the Granary: Observations on Nineteenth Century Medical Narratives.' In: *Literature and Medicine* 25/2, pp. 255–276.
Sacklén, Karl Wilhelm (1870): Om Transfusion, Helsingfors: Frenckell & Son.
Sahlins, Peter (2015): 'The Beast Within: Animals in the First Xenotransfusion Experiments in France, ca. 1667–68.' In: *Representations* 129/1, pp. 25–55.
Sander, Friedrich (1874): 'Zur Lammbluttransfusion.' In: *Berliner Klinische Wochenschrift* vol. 11, pp. 173–175, 189–191.
Sanderson, Burdon (1873): Handbook for the Physiological Laboratory, London: J. A. Churchill.
Scheel, Paul (1802): Die Transfusion des Blutes und Einsprützung der Arzeneyen in die Adern, Copenhagen: Friedrich Brummer.
Schivardi, Plinio (1874): 'La trasfusione del sangue neigli alienati.' In: *Gazetta Medica Italiana* vol. XXXIV, pp. 289–291, 297–299.
Schlich, Thomas (2004): 'The Emergence of Modern Surgery.' In: Deborah Brunton (ed.), Medicine Transformed. Health, Disease and Society in Europe 1800–1930, Manchester: Open University, pp. 61–91.
Schlich, Thomas/ Mykhalovskiy, Eric/ Rock, Melanie (2009): 'Animals in Surgery – Surgery in Animals; Nature and Culture in Animal-human Re-

lationship and Modern Surgery.' In: *History and Philosophy of the Life Sciences* vol. 31, pp. 321–354.

Schliep, Paul (1874): 'Fall von directer arterieller Tierblut-Tranfusion.' In: *Berliner klinische Wochenschrift* vol. 11, pp. 25–28.

Schmidt, E. (1874): 'Zur Transfusion bei chronischen Krankheiten.' In: *Aerztliche Mittheilungen aus Baden* vol. XXVIII, pp. 137–141, 147–151.

Schmidt, E. (1875): 'Endresultat von 8 Lammbluttransfusionen.' In: *Aerztliche Mittheilungen aus Baden* vol. XXIX, pp. 106–108.

Schneider, William H. (1997): 'Blood Transfusion in Peace and War, 1900–1918.' In: *Social History of Medicine* 10/1, pp. 105–126.

Schorr, Marianne (1956): Zur Geschichte der Bluttransfusion im 19. Jahrhundert. Unter besonderer Berücksichtigung ihrer biologischen Grundlagen, Basel: Benno Schwabe.

Severn, Charles (ed.)(1839): James Blundell's Lectures on the Principles and Practice of Midwifery, London: J. Masters.

[Seyfer]th (1874): 'Über die Lammblut-Transfusionen.' In: *Correspondenzblätter des Allgemeinen ärztlichen Vereins von Thüringen* April 15, pp. 43–48.

Showalter, Dennis (2004): The Wars of German Unification, London: Arnold.

Simon, Jonathan (2007): 'The Origin of the Production of Diphtheria Antitoxin in France, between Philanthropy and Commerce.' In: *Dynamis* 27, pp. 63–82.

Sittel, Theo (1874): 'Lamb's blood transfusion.' In: *The Clinic* vol. VI, pp. 157–158, 169–170, 183–184.

Stapelberg, Monica-Maria (2016): Through the Darkness. Glimpses into the History of Modern Medicine, Crux Publishing.

Starr, Douglas (1998): Blood. An Epic History of Medicine and Commerce, New York: Alfred A. Knopf.

Statistisches Jahrbuch, Stuttgart, 1963.

Steiner, Franz (1874): 'Zwei Thierbluttranfusionen nach einer Amputation des Oberschenkels.' In: *Wiener Medizinische Wochenschrift* vol. 24, pp. 308–311.

Stern, H.A. (1874): Drei Lammbluttransfusionen bei Phthisikern, Halle: Plötz'sche Buchdrückeri.

Stöckel, Sigrid/Lisner, Wiebke/Rüve, Gerlind (2012): Das Medium Wissenschaftszeitschrift seit dem 19. Jahrhundert, Stuttgart: Franz Steiner Verlag.

Stockholms Korrespondens, June 10, 1875

Stowe, Steven M. (1996): 'Seeing Themselves at Work: Physicians and the Case Narrative in the Mid-Nineteenth-Century American South.' In: *The American Historical Review* 101/1, pp. 41–79.

Sutherland, A. J. (1855): 'Cases illustrating the Pathology of Mania and Dementia.' In: *Medico-chirurgical transactions* 38, pp. 261–288.

Svenska läkaresällskapets förhandlingar (1875) In: *Hygiea* vol. 37, pp. 28.

Svensson, Ivar (1875): 'Transfusion två gånger utförd på samma patient.' In: *Uppsala Läkareförenings förhandlingar* vol.10, pp. 329–337.

Swain, Valentine E. J. (1970): 'Franco-Prussian War 1870-1871: Voluntary Aid for the Wounded and Sick.' In: *British Medical Journal* vol. 29, pp. 514.

Tagliavini, Annamaria (1985): 'Aspects of the History of Psychiatry in Italy in the Second Half of the Nineteenth Century.' In: W.E. Bynum/Roy Porter/Michael Shepherd (eds.), The Anatomy of Madness. Essays in the History of Psychiatry, London/New York: Tavistock, pp. 175–196.

Tamburini, Augusto (1874): 'La trasfusione del sangue nella pellagra.' In: *Lo Sperimentale* vol. XXXIV, pp.186–196.

The Doctor (1875), pp. 144.

The Journal of Mental Science (1876), pp. 11–12.

The Lancet, September 26 and October 10, 1874.

The Obstetrical Journal of Great Britain and Ireland (1873), pp. 360.

The Obstetrical Journal of Great Britain and Ireland (1873/74), pp. 543.

Thurn, Dr (1874a): 'Directe Lammbluttransfusion.' In: *Berliner klinische Wochenschrift* vol. 11, pp. 393–395.

Thurn, Dr (1874b): 'Nachtrag zu den in no 32 1874 mitgeteilten Fällen von Lammblut-Transfusion.' In: *Berliner klinische Wochenschrift* vol. 11, pp. 658–659.

Tillhagen, Carl-Herman (1989): Vår kropp i folktron, Stockholm: Gummesson.

Torstensson, Otto (1874a): Några underrättelser om Mössebergs vattenkuranstalt, Göteborg: J.A. Granberg.

Torstensson, Otto (1874b): 'Lammblodstransfusion.' In: *Hygiea* vol. 36, pp. 674–677.

Transactions of the Medical society of North Carolina (1872), pp. 40–45.

Trendelenburg, Friedrich (2011): Die ersten 25 Jahren der Deutschen Gesellschaft für Chirurgie, Hamburg: Severus.

Tschin, Dr (1874): 'Eine chinesische Transfusionsgeschichte.' In: *Wiener Medizinische Presse* vol. XV, pp. 1169–1170.

Tucker, G.A. (1887): Lunacy in Many Lands, Sydney: Charles Potter.

Tucker, Holly (2011): Blood Work. A Tale of Medicine and Murder in the Scientific Revolution, New York/London: Norton.
v. Klebelsberg, Ernst (1922): 'Tierbluteinspritzungen bei Psychosen.' In: *Zeitschrift für die gesamte Neurologie und Psychiatrie* vol.76, pp. 611–626.
Verga, Andrea (1875): 'Introduzione.' In: *Archivio Italiano per le Malattie nervose e particolarmente per le Alienazioni mentali* vol. 12, pp. 415–418.
Vizioli, Francesco (1874): 'Intorno un altro caso di trasfusione diretta di sangue in donna anemica.' In: *Il Morgagni* vol. XVII, pp. 673–685.
Vizioli, Francesco (1875): 'Intorna la trasfusione di sangue.' In: *Il Morgagni* vol. XVII, pp. 279–285.
Volz, Robert (1870): Der ärztliche Beruf, Berlin: M.C. Charitius.
von Belina-Swiontkowsky, L. (1869): Die Transfusion des Blutes in physiologischer und medicinischer Beziehung, Heidelberg: Carl Winter.
von Bergmann, Ernst (1883): Die Schicksale der Transfusion im letzten Decennium, Berlin: Hirschwald.
von Cube, Dr (1877): 'Zur Casuistik der directen Thierbluttransfusion.' In: *Deutsche Medicinische Wochenschrift* vol. 3, pp. 378–381.
von Esmarch, Friedrich (1894): Handbuch der Kriegschirurgischen Technik, Kiel/Leipzig: Lipsius & Tischer.
von Linhart, Wenzel (1874): Compendium der chirurgischen Operationslehre, Wien: Wilhelm Braumüller.
von Weber, Ernst (1879): *Videnskabens torturkamre: en samling af kjendsgjerninger*, København: Foreningen til dyrenes beskydelse.
von Weber, Ernst (1880): *Vetenskapens tortyrkamrar*, Stockholm: Looström & K.
Waldenström, J. A. (1875): 'Transfusion med anledning af blödning efter förlossning.' In: *Upsala Läkareförenings Förhandlingar* vol.10, pp. 191–196.
Wallis, Curt (1876): 'Den direkta lamblodstranfusionens användande mot lungsot.' In: *Hygiea* vol. 38, pp. 15–37.
Warfvinge, F. W. (1876): 'Öfverblick af transfusionsfrågan jemte redogörelse för några transfusioner verkstälda II.' In: *Hygiea* vol. 38, pp. 132–153.
Warfvinge, F. W. (1878): 'Meddelande vid Svenska Läkare-sällskapets sammankomst 16 juli 1878, In: *Svenska Läkare-sällskapets Förhandlingar*, pp. 133–140.
Wawro, Geoffrey (1996): The Austro-Prussian War. Austria's War with Prussia and Italy in 1866, Cambridge: Cambridge University Press.
Wawro, Geoffrey (2000): Warfare and Society in Europe 1792–1914, London/New York: Routledge.
Westerberg, A. P. (1875): 'Fall af lammblodstransfusion vid långt framskriden lungsot.' In: *Hygiea* vol. 37, pp. 139–141.

Westergren, Georg E. (1880): 'Om transfusionens framtid inom den militära chirurgien,' In: *Tidskrift i militär helsovård* vol.3, pp. 59–67.

Whitaker, Elizabeth D. (1992): 'Bread and Work: Pellagra and Economic Transformation in Turn-of-the-Century Italy.' In: *Anthropological Quarterly* 65/2, pp. 80–90.

Wiener Medizinische Wochenschrift (1874), pp. 1094.

Wilde, Sally (2009): 'Truth, Trust, and Confidence in Surgery, 1890–1910: Patient Autonomy, Communication, and Consent.' In: *Bulletin of the History of Medicine* 83/2, pp. 302–330.

Winants, J.E. (1872): 'Case of Transfusion.' In: *The American Journal of the Medical Sciences* vol. LXVI, pp. 108.

Worm-Müller, Jakob (1875): Transfusion und Plethora. Eine physiologische Studie, Christiania: W.C. Fabritius.

Wurtz, Adolphe (1870): Les hautes études pratiqués dans les universités allemandes, Paris : Imprimerie Impériale.

Zannetti, R. (1876): 'Sulle memorie e opere inviate alla Società medico-fisica fiorentini durante l'anno 1875.' In: *Atti dell'Accademia medico-fisica Fiorentina*, pp. 77–79.

Zeitschrift für Bauwesen (1861), pp. 53.

Zimmer, Arnold/ v. Balden, Eleonore (1931): 'Die Tierblutbehandlung der Basedowschen Krankheit.' In: *Deutsche Zeitschrift für Chirurgie* vol. 234, pp. 236–244.

Zimmerman, Leo M./Howell, Katherine M. (1932): 'History of Blood Transfusion.' In: *Annals of Medical History* vol. IV, pp. 415–433.

Acknowledgements

The lamb blood transfusions, chronicled in this book, occurred across Europe and the USA in the late 19th century. The events were reported in medical journals, newspapers and magazines in many different languages. My 21st century investigation into this strange phenomenon owes a lot to the anonymous workers in libraries and elsewhere who have digitalized these old pages and made them available for research.

Still, I would not have found some crucial material without the competent help of archivists in several European countries. My heartfelt thanks, especially, to Dr Chiara Bombardini, Director of Library and Archive of S. Lazzaro Asylum in Reggio Emilia, for her skilled and generous assistance. Thanks also to Dr Dirk Alvermann at Universitätsarchiv Greifswald and Dr Wolfram G. Theilemann at the Nordhausen Stadtarchiv, to Professor Nils Mandahl for searches in the Swedish National Archives, and to The Evangelisches Kirchspiel am Auerberg for information from the Kirchenbücher in Schwenda. I have received good help from archivists at the Red Cross in Geneva and the Obstetrical Society of London. Special thanks to Francesca Poli for an insightful tour of the Museo di Storia della Psichiatria in Reggio Emilia.

I wish to thank the many colleagues who have commented on previous versions presented at seminars and conferences in Sweden and elsewhere. Elisabeth Hedborg provided invaluable help in deciphering the poetry and handwritten prose of the Italian alienists – lots of thanks for your enthusiasm and encouragement! Many thanks, too, to David Lindberg for crucial assistance with the images. Bengt Olle Bengtsson has accompanied me during all stages of the journey towards a completed book. It owes a lot to his patience, generosity and constructive criticism. I thank him with all my heart.

The publication of this book was made possible by generous grants from Gunvor och Josef Anérs Stiftelse, Åke Wibergs Stiftelse, Stiftelsen Lars Hiertas Minne, and Linköping University, for which I am grateful.

Index of Places

A
Addison, 63, 69
Alessandria, 81, 92, 93, 102, 134, 135
Australia, 161
Austria, 41, 47, 48, 57, 59, 88, 136, 139, 140, 146, 154, 163

B
Barmen, 17, 64
Belgium, 155
Berlin, 32, 33, 34, 41, 51, 53, 58, 64, 114, 139, 147, 153, 161, 162, 163
Bologna, 88
Bordeaux, 119, 164, 174n
Boston, 64, 71, 136
Brescia, 81, 95, 96-101, 135, 153, 176n
Breslau, 34, 57
Bulgaria, 146

C
Champigny, 51
Chile, 151, 172n
China, 46
Cincinnati, 64, 135

Cologne, 64
Copenhagen, 107, 109, 110, 153
Custoza, 47

D
Dalston, 65
Denmark, 32, 47, 49
Dresden, 66, 77, 121
Düren, 72

E
Edinburgh, 21, 22
Elvira, 64
Emilia, 82, 86
England, 19, 20, 23, 65, 89, 137, 139, 151, 167n

F
Falköping, 62
Fall River, 64
France, 19, 38, 41, 45, 47, 82, 89, 114, 130, 159–161, 164, 167n

G
Gadebusch, 64
Geneva, 36, 37
Genoa, 146

Germany, 9, 17, 31, 42, 49, 51, 59, 63, 64, 105, 108, 110, 146, 156, 161, 174n
Gotland, 145
Great Britain, 23, 26, 65, 114, 168n, 178n, 180n
Greifswald, 32, 34, 105, 106, 114, 153, 162, 169n, 173n, 183n, 190n

H
Hall, 163
Harz, 9, 11, 32, 62, 153
Heidelberg, 114

I
Illinois, 63, 69
Imola, 81, 88, 90, 92, 94, 95, 101
Iowa, 147
Italy, 11, 18, 28, 29, 47, 81-101, 110, 139, 146

J
Jena, 71

K
Karlsruhe, 64
Kiel, 109
Kirchbarkau, 172n
Königgrätz, 48, 49
Kyrkefalla, 62

L
Lahr, 64
Leipzig, 114

Lombardy, 82
London, 11, 20, 21, 65

M
Mainz, 64
Malvicino, 92
Massachusetts, 64
Mecklenburg, 68
Mexico, 46
Milan, 53, 85, 95
Modena, 91
Mösseberg, 61-62, 66, 69–73, 165
Munich, 122

N
Nantes, 161
Naples, 28, 89, 92
New York, 137
New Zealand, 66, 161
Niederrad, 64
Nordhausen, 9, 11, 31, 32, 33, 35, 49, 62, 69, 107, 108, 153, 154
North Carolina, 11, 28, 29

O
Ohio, 64
Orléans, 55
Oskarshamn, 121

P
Padua, 18
Paraguay, 46

Paris, 20, 26, 36, 38, 49, 51,
    102, 109, 112, 132, 159,
    161, 167n, 170n, 172n,
    174n
Pavia, 81, 95
Pesaro, 81, 94
Pforzheim, 71
Pontarlier, 45
Prussia, 47

Q
Quedlinburg, 32

R
Reggio Emilia, 81, 85, 86, 87, 91,
    92, 115, 165, 175n, 176n,
    182n
Rome, 88
Rostock, 106, 116
Russia, 35, 36, 39, 41, 59, 154

S
Sadowa, 48
Scandinavia, 151
Schleswig-Holstein, 172n
Schwenda, 9, 30, 165
Solferino, 47
St. Petersburg, 11, 29, 32, 34,
    35–36, 39, 64, 65, 110,
    140, 153
Stockholm, 64
Sweden, 11, 61, 62, 66, 68, 73, 145,
    146, 165n, 175n
Switzerland, 36, 45

T
Turin, 134, 135

U
USA, 11, 12, 30, 31, 61, 63, 72, 111,
    150, 151, 161

V
Valparaiso, 172
Västergötland, 62, 165
Veneto, 82
Verona, 48
Verrières, 45
Vienna, 39, 64, 86, 114, 139, 175n
Vittlånge, 145

W
Wiesbaden, 32
Wilmington, 28
Würzburg, 109

Index of Names

A
Albini, 28, 29, 53, 65, 89, 92, 168n, 171n
Aranda, 113
Armstrong, 179n
Aveling, 23, 65, 115, 118, 133, 172n

B
Balden von, 163, 182n
Barnes, J., 170n
Barnes, R., 133, 179n
Bartrip, 169n
Beck, 170n
Behring, 161
Beigel, 173n
Belina von, 122, 179n
Bergmann von, 147-148, 153, 156, 162, 181n
Bergonzi, 177n, 180n
Bernard, 102, 109, 129, 138, 155, 180n
Berner, 167n
Bernheim, 159–161, 182n
Bertin, 161
Bier, 162-163, 182n
Billroth, 41, 168n, 169n, 170n
Birch-Hirschfeld, 75, 77-78, 121, 172n, 174n, 178n
Birke, 179n
Björck, 69–71, 173n, 179n
Björnström, 167n, 179n
Bleker, 177n
Blessich, 94
Blessing, 182n
Bliedung, 172n
Blundell, 21–22, 23, 26, 27, 89
Bonacossa, 134, 135
Bozzolo, 177n, 179n
Braun, 169, 173n
Bretschneider, 179n
Briggar, 64
Brocca, 85, 175n
Brown, 179n
Bruberger, 51, 57, 171n, 180n
Brügelmann, 64, 76, 172n, 174n
Buess, 169n
Bynum, H., 172n
Bynum, W., 177n, 180n
Byron, 22

C
Caretti, 91
Cascella, 176n
Caselli, 92
Casse, 25, 168n
Cesalpino, 89
Chadwick, 136, 169n, 173n, 180n
Chalmers, 178n

Chenu, 171n
Christina, 20
Cooter, 58, 171n
Cradle, 173n
Cristoforis de, 95, 134, 175n, 176n, 179n, 180n,
Cruchet, 164, 182n
Cube von, 75, 120, 121, 174n, 178n

D
Dallera, 146, 181n
Danic, 170n
Daniel D., 169n, 181n
Daniel T., 172n
Delorme, 156, 182n
Denis, 20, 89, 110, 121, 132
Descola, 132, 179n
Dessauer, 172n
Dieffenbach, 21, 27, 168n
Domenicis, 161
Douglas, 133, 179n
Dumas, 23, 27
Dyck, 180n
Dziewonski, 170n, 181n

E
Eckert, 53–55, 56, 171n, 179n
Edholm, 171n
Elkeles, 152, 181n
Engdahl, 180n
Esmarch von, 53, 157, 182n
Eyselein, 180n

F
Faber, 172n
Famius, 164, 182n
Fazio de, 175n
Fennell, 180n
Fieber, 139- 140

Fiedler, 75, 77-78, 121, 172n, 174n, 178n
Fildes, 109
Fischer, H., 57, 170n, 171n
Fischer, K., 170n
Flemming, 64, 68, 172n, 173n
Folli, 18
Forbes, 170n

G
Gabriel, 170n
Galileo, 102
Garrison, 170n
Gazzaniga, 194
Geissler, 33
Gentilcore, 175n
Gesellius, 13, 14, 24, 29–32, 34–36, 40, 41, 52–54, 57, 66, 89, 107, 109, 110, 122, 133, 139, 153, 168n, 169n, 171n, 173n, 179n
Gilder, 167n
Ginnaio, 175n
Girerd, 181n
Gissler, 71, 173n
Goethe, 17, 148
Gould, 179n
Greenwalt, 168n
Grosberg, 169n, 181n
Guerrini, 12, 167n, 179n, 180n

H
Hadenfeldt, 163, 182n
Haeckel, 129
Haehner, 181n
Halpern, 180n
Harvey, 18, 89, 129

Hasse, 9, 11, 13, 14, 29–36, 40–42, 49–51, 53, 57, 58, 62–66, 68, 69, 71–78, 86, 89, 108–110, 116, 121, 123, 129, 134, 136, 139, 140, 149, 153, 154, 162, 165n, 167–174n, 178n, 179n, 181n,
Hayem, 148, 149, 153, 155, 181n
Hertzberg, 157, 182n
Hess, 170, 190
Heyfelder, 35, 36, 39, 53, 64, 72, 77, 154, 169n, 171–174n, 179n,
Hippocrates, 148
Hirschfeld, 179n
Hjert, 145
Hobsbawm, 170n
Hoffman, 64
Hotz, 63, 64, 172n
Howell, 167n, 175n, 181n
Hueter, 173n
Hurwitz, 173n

I
Icard, 171n
Iorio, 194

J
Jacobsson, 62, 70, 72, 165
Jahn, 119, 122, 178n, 179n
Jenkins, 181n
Jenner, 138
Jennings, 150, 173n, 181n
Jewson, 177n

Jullien, 107, 120, 170n, 175n, 177n
Jürgensen, 168n

K
Kaufmann, 18
King, E., 20, 89
King, Dr, 28
Kitasato, 161
Klebelsberg von, 163, 182n
Klingelhöffer, 64, 68, 172n, 173n
Koch, 63, 159
Köhler, 167, 170n, 181n, 182n
Krüger, 9- 10, 30, 165
Küster, 64, 172n

L
Laguzzi, 81, 92, 174n
Lahaie, 182n
Landois, 39, 105-106, 115–118, 121, 123-124, 129–131, 140, 147, 153, 162, 163, 167–169n, 177–180n
Landsteiner, 106
Larsson, 118
Leacock, 21
Lefrère, 170n
Leidesdorf, 86, 175n
Leisrink, 25, 168n, 169n
Lesser, 114, 121, 178n
Leube, 71, 174n
Lidell, 48, 170n
Lindes, 178n
Linhart von, 178n
Lisner, 199

Livi, 85, 86, 91-92, 100, 102, 153, 175–177n
Ljungström, 181n
Logan, 179n
Lombroso, 83, 84, 95, 124, 153, 175–177n, 179n
Lovén, 178n
Lower, 19, 20
Luigi, 176n

M
Madge, 133, 134, 172n, 180n
Maehle, 135, 136, 180n
Magendie, 27, 129, 149
Maluf, 167n
Manfredi, 29
Mantegazza, 130, 139, 177n, 179n
Manzini, 95–102, 121, 129, 140, 153, 175- 177n, 179n
Mariani-Costantini, 175n
Marinozzi, 167n, 168n, 175n
Marmonnier, 26, 122, 168n, 179n
Martin, 122, 168n, 179n
Masing, 75, 174n, 179n
McCallum, 170n
Merkel, 64, 71, 74, 172–174n
Meyer, 160
Michael, 179n
Michetti, 94, 176n
Molitor, 64, 172n
Moore, 167n
Morselli, 139, 175, 177, 178n, 180n
Morton, 172n
Müller, 168n, 177n, 181n

Mykhalovskiy, 198

N
Neudörfer, 39, 48, 54, 56, 57, 64, 67, 69, 86, 117, 129, 136, 154, 169–173n, 178–180n,
Niemeyer, 38, 46, 170n
Nussbaum von, 55, 171n

O
Ockert, 69
Olivier, 169, 181n
Oré, 119, 121–123, 164, 168n, 171n, 174n, 178n, 179n
Ovid, 17

P
Paglierani, 94
Panum, 24, 27, 107, 109–111, 116, 121, 123, 124, 127–131, 139, 147, 149, 153, 162, 168n, 169n, 174n, 177–181n,
Pasteur, 159
Pelis, 168n, 172n, 178n, 181n, 182n
Pellegrino, 180n
Penzoldt, 169n, 174n
Pepys, 20
Petersson, 173n
Pettersson, 181n
Picq, 161
Polidori, 22
Polli, 95, 176n
Ponfick, 106, 116, 118, 119, 147, 153, 163, 177n, 178n

Porro, 176n
Prévost, 23, 27
Priani, 175n
Purmann, 18, 19

R
Rader, 179n
Ransanigo, 98
Raz, 177n
Redtel, 74, 76, 174n, 177n
Reeder, 175n
Rémy, 179n
Richards, 178n, 179n
Richet, 161
Riesman, 168n
Risse, 178n
Riva, 18, 89
Rock, 198
Rodolfi, 95–102, 121, 129, 140, 153, 175-177n, 179n
Roelen, 72, 173n, 179n
Röhlen, 174n
Rosa, 89
Rosenberg, 178n
Rossander, 116, 119, 178n, 181n
Rossi, 165
Roth, 170n
Roussel, J., 13, 14, 23, 30, 31, 35–42, 45–48, 51, 52, 55–57, 65, 71, 118, 124, 154–156, 159, 168–171n, 175n, 178n, 179n, 181n, 182n
Roussel, T., 175n
Ruffini, 98

Rüve, 199n
Rylance, 173n

S
Sacklén, 122, 168n, 179n
Sahlins, 179n
Sander, 17, 21, 64, 167n, 172n
Sanderson, 178n
Santi de, 170n, 171n
Scheel, 20, 167n, 168n
Schivardi, 175n, 177n, 179n, 180n
Schlich, 178n, 179n
Schliep, 51, 53, 57, 64, 71, 171n
Schmidt, 64, 140, 172n, 180n
Schneider, 182n
Schorr, 168n, 178n
Scipion, 172n
Selmi, 86-87
Severn, 168n
Shelley, 21, 22
Showalter, 170n
Simon, 182n
Sittel, 64, 136, 167n, 172n, 180n
Stapelberg, 181n
Starr, 167n, 168n, 177n
Steiner, 180n
Stern, 76-77, 174n
Stewart, 180n
Stöckel, 169n
Stowe, 173n
Sutherland, 90, 175n
Svensson, 77, 121, 135, 173n, 174n, 178n, 180n
Swain, 170n

T

Tagliavini, 174n, 175n, 177n
Tamburini, 86, 175n, 177n
Tector, 145
Thomasma, 180n
Thurn, 64, 77, 172n, 174n
Tillhagen, 181n
Torstensson, 61–63, 71, 171n, 172n, 173n
Trendelenburg, 180n
Tschin, 140, 180n
Tucker, G.A., 174n, 175n
Tucker, H., 167n

V

Verga, 95, 177n
Vernia, 176n
Virchow, 109, 112
Vizioli, 175n, 180n
Voisin, 174n
Volta, 176n
Volz, 178n

W

Waldenström, 173n
Wallis, 78, 107, 124, 172n, 174n, 177n, 179n
Warfvinge, 64, 116, 123-124, 172n, 173n, 178n, 181n
Watier, 182n
Wawro, 170n
Weber von, 127, 131, 179n
Wentzel, 71, 173n
Westerberg, 136, 172n, 180n
Westergren, 169n
Weyland, 64
Whitaker, 175n
Wilde, 180n
Williams, 130, 179n

Winants, 28, 168n
Worm-Müller, 106, 177n
Wurtz, 178n

Z

Zannetti, 177n
Ziemssen, 66, 191n
Zimmer, 163, 182n
Zimmermann, 167n, 175n, 181n

Historical Sciences

Federico Buccellati, Sebastian Hageneuer,
Sylva van der Heyden, Felix Levenson (eds.)
Size Matters – Understanding Monumentality Across Ancient Civilizations

2019, 350 p., pb., col. ill.
44,99 € (DE), 978-3-8376-4538-5
E-Book: available as free open access publication
E-Book: ISBN 978-3-8394-4538-9

Jochen Althoff, Dominik Berrens, Tanja Pommerening (eds.)
Finding, Inheriting or Borrowing?
The Construction and Transfer of Knowledge in Antiquity and the Middle Ages

2019, 408 p., pb., ill.
54,99 € (DE), 978-3-8376-4236-0
E-Book: available as free open access publication
E-Book: ISBN 978-3-8394-4236-4

Judith Mengler, Kristina Müller-Bongard (eds.)
Doing Cultural History
Insights, Innovations, Impulses

2018, 198 p., pb., col. ill.
34,99 € (DE), 978-3-8376-4535-4
E-Book: 34,99 € (DE), ISBN 978-3-8394-4535-8

All print, e-book and open access versions of the titles in our list
are available in our online shop www.transcript-verlag.de/en!

Historical Sciences

Laura Meneghello
Jacob Moleschott – A Transnational Biography
Science, Politics, and Popularization
in Nineteenth-Century Europe

2017, 490 p., pb.
49,99 € (DE), 978-3-8376-3970-4
E-Book: 49,99 € (DE), ISBN 978-3-8394-3970-8

Johnny Van Hove
Congoism
Congo Discourses in the United States
from 1800 to the Present

2017, 360 p., pb., ill.
39,99 € (DE), 978-3-8376-4037-3
E-Book: available as free open access publication
E-Book: ISBN 978-3-8394-4037-7

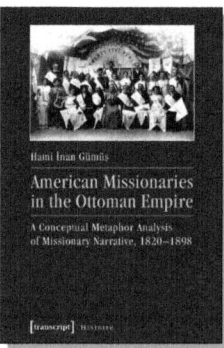

Hami Inan Gümüs
American Missionaries in the Ottoman Empire
A Conceptual Metaphor Analysis
of Missionary Narrative, 1820-1898

2017, 260 p., pb.
34,99 € (DE), 978-3-8376-3808-0
E-Book: 34,99 € (DE), ISBN 978-3-8394-3808-4

All print, e-book and open access versions of the titles in our list
are available in our online shop www.transcript-verlag.de/en!